Ask an Ocean Explorer

Dr Jon Copley has more than twenty years of experience in exploring the deep ocean and he has been a scientific advisor for TV series including *Blue Planet II*. He frequently talks about exploring the deep ocean to public audiences ranging from local schools and community groups to major events and festivals.

Ask an Ocean Explorer

Jon Copley

HODDER

First published in Great Britain in 2019 by Hodder & Stoughton
An Hachette UK company

This paperback edition published in 2019

1

Copyright © Jon Copley 2019

Paperback ISBN 9781473696907
eBook ISBN 9781473696884

Typeset in Dante MT by Hewer Text UK Ltd, Edinburgh
Printed and bound in Great Britain by Clays Ltd, Elcograf S.p.A.

Hodder & Stoughton policy is to use papers that are natural, renewable and recyclable products and made from wood grown in sustainable forests. The logging and manufacturing processes are expected to conform to the environmental regulations of the country of origin.

Hodder & Stoughton Ltd
Carmelite House
50 Victoria Embankment
London EC4Y 0DZ

www.hodder.co.uk

For Claire and Zena

Introduction

When we think of the ocean, perhaps we imagine turquoise waters lapping on a pale sandy shore somewhere in the tropics, with dolphins sporting through the dappling waves offshore. And if we think of the deep ocean, maybe we imagine a mysterious place cloaked in darkness and crushing pressures, inhabited by monstrous squid and strange species resembling creatures from science fiction.

It is all of those things, and much more.

The ocean gives our world its 'blue planet' appearance from space, yet if the Earth were the size of a football, all of the water in the ocean would only be as thin as a sheet of paper wrapped over its surface. The area and depth of the ocean is vast compared with the scale of an individual human, but our collective activities as a species are affecting its furthest recesses. Most of the species living in the ocean lack backbones, unlike its iconic whales, dolphins, sharks and turtles, and we are still discovering new species of animals in the deep right now.

I have spent most of my career trying to grasp the reality of our blue planet, as an 'ocean explorer', and I hope to share some of that reality with you in this book. But 'ocean explorer' is not a job that I saw advertised somewhere and applied for; it's a generous description for the overall direction of a wandering path. So by way of an introduction, I'll try to answer a question that I am sometimes asked: 'how do you become an ocean explorer?'. The short answer is 'by becoming a scientist', and perhaps that path begins sooner than we realise. I think we're all born scientists, with a natural curiosity about our world, and sometimes experiences or elements around us feed that further.

I learned to swim late, at about age eight, because I was scared of the water. I didn't like the feeling of being out of my depth without a float, and I didn't like the idea of what might be lurking out-of-sight underwater. But what changed that for me was trying on a diver's face-mask and being able to peer into a startling clear volume beneath the surface. Having learned to swim, I was then less interested in perfecting strokes to travel quickly at the surface, and much more interested in swimming along the bottom of the pool, spending as long as I could submerged. I had my first try-out session of scuba diving as a young teenager, taught by a former oil rig diver, and loved the three-dimensional freedom of it, being able to rise and sink just by breathing in and out, and moving through a liquid space so different from our everyday world.

Meanwhile, I grew up fascinated by animals, surrounded by a menagerie of family pets, and inspired by TV wildlife documentaries. A local stream and ponds provided my first opportunity to explore underwater life, with a dipping net and a jam-jar. Mosquito larvae in particular caught my imagination, living underwater and breathing through a tube at the tip of their tail, with bodies so different to the familiar adult insect. Occasionally I'd come across a dragonfly nymph too and be in awe of its telescopic jaws – an amazing underwater hunter, which would be terrifying to us if we were the same size as its prey. The notion that there could be equally weird and wonderful species still waiting to be discovered out there perhaps lodged somewhere at that point.

For many years while growing up I wanted to become a veterinary surgeon, but eventually I realised that I was more interested in what animals do, rather than learning to treat them for illness or injury. After some teenage vacillations, I applied to read zoology at university, which included some marine biology but also covered life on land. One day I was wandering through the university library when I saw a book on a shelf with a picture of some odd-looking underwater animals on its cover: a cluster of red-plumed wormish creatures, living in stark white tubes somewhere in the dark of the deep ocean. Despite being halfway through a zoology degree, I couldn't put a name to those animals, not even to the broad category of animal life to which they might belong. So I fetched down the volume, which turned out to be a collection of research papers about newly discovered life at hot springs on the ocean floor, known as hydrothermal vents.

Reading through it was a revelation. At school, I had been taught that all life depended on the sun as an ultimate source of energy: plants use sunlight to grow, and in turn provide food for herbivores, which are themselves eaten by carnivores, and so on. But the life around these deep-sea hydrothermal vents is sustained by chemical energy: minerals in the hot fluids gushing out of the ocean floor nourish microbes, which in turn provide food for the animals, including the worm-like creatures that caught my eye. None of this had been mentioned in my lecture classes, because this deep-sea discovery was so new, and yet it rewrote the rules of the textbooks.

From that point on, deep-sea hydrothermal vents became an obsession for me, and whenever I could twist the title of an assignment to find out more about them, I did. At the end of my degree, my tutor advised me not to pursue a PhD – the equivalent of an apprenticeship in science, through which you become an independent investigator – unless it was to carry out research on something that captivated me utterly, for which I would be prepared to give blood, sweat and tears. Out of all the subjects I had touched on in my degree, only one fitted the bill: those deep-sea hydrothermal vents. But at the time, there were no biologists in the UK exploring them.

I kicked my heels for a while, working to earn money for further study, and then took a Master's degree in oceanography, which enabled me to specialise in marine biology and introduced me to the physics, chemistry and geology of the oceans – a programme of classes that I enjoyed thoroughly, and have since enjoyed teaching years later. During my master's at the University of Southampton, my tutor, Dr Paul Tyler, obtained the first research grant in the UK to investigate the biology of deep-sea hydrothermal vents, with funding for a research assistant who would be able to undertake some of the work for a PhD. That was me, and five years after blundering across deep-sea hydrothermal vents in a library, I ended up aboard a Russian research ship heading out to explore them in the Atlantic, and a year after that I was visiting them in a US Navy mini-submarine at the bottom of the Pacific.

With hindsight we can link events to make seemingly tidy narratives, but in reality, the path that we follow involves some things that happen by chance. If I hadn't been in the right place at the right time for a project investigating hydrothermal vents, I might have done something else with my life. My work since my PhD has occasionally meandered away from ocean exploration, in between the temporary positions and bouts of research funding that chart a scientific career; I've also been a science journalist and co-founded a company that trains scientists in how to communicate their work beyond their specialist colleagues. But always, there's the pull of the unknown embodied in the deep ocean – and always just one more project or expedition to propose, to reach the next unexplored vista.

Over the past twenty-five years, my research has explored places such as hydrothermal vents in the Atlantic, Pacific, Indian and Southern Oceans; the seafloor earthquake zone that triggered the Boxing Day tsunami of 2004; hotspots of deep-sea life in the Gulf of Mexico; and underwater mountains around Antarctica. I have lived aboard research ships of different nations, working at sea with colleagues whose backgrounds range from geophysics to engineering. Together we have seen previously hidden parts of our planet, and encountered species that no-one had seen before, figuring out the puzzles of how they thrive in the deep ocean.

The biggest change that I have noticed during the past quarter-century has been the motivations for our voyages. When I started out, deep-sea biology was largely a branch of 'pure' research, with a goal of obtaining new knowledge about our world. But the work has become much more 'applied', motivated by the need to understand the impacts that our lives are having on the world, even in its deep oceans. I now spend more of my time trying to communicate with politicians and policy-makers about what we are finding, to help them make more informed decisions about activities that affect the deep ocean. In recent years I have also worked with TV, radio, and online documentary-makers to share the exploration of the deep ocean with people worldwide, and interacted with news media to raise awareness of deep-sea issues and my team's discoveries. It's no longer enough just to publish our findings for colleagues to digest in scientific journals, though of course we still do that so that our conclusions are thoroughly checked by other scientists. The deep ocean covers most of our world, and that's too important to be left just to scientists and professional policy-makers: I'd like more of us to be aware of how our own lives are connected to the hidden half of our planet.

That's why I have written this book: what we now know about the deep ocean is as astonishing as the unknown that remains, and we all need to be involved in the choices for its future. But everything we know and don't yet know about the deep ocean cannot possibly fit into a volume like this one, so this book is really just a dipping net and a jam-jar, and I hope that the inevitably eclectic collection in these pages rekindles some of the childhood curiosity of exploring with those tools. To me, the real wealth of the deep ocean isn't in the mineral deposits being chased by seafloor miners, or even the spin-off benefits of new materials and medical treatments that come from studying the species that live down there. It's the capacity of this frontier to renew our sense of wonder about the world around us, allowing us to see it again as if for the first time.

I

How deep are the oceans and how do we explore them?

In December 1968, astronauts aboard *Apollo 8* on the way to orbit the Moon became the first people to take photographs of the entire globe of the Earth from space. The expanse of blue in those pictures, interrupted by swirling clouds and the occasional continent, vividly conveys how the oceans cover 71 per cent of our planet. But what those pictures cannot illustrate is just how deep those oceans are. In 1521 the explorer Ferdinand Magellan declared the ocean to be 'immeasurably deep' when he reputedly lowered a cannonball on 700 metres of rope beneath his ship in the Pacific and it showed no signs of touching the seabed. But we now know that the deepest point in the oceans is the 'Challenger Deep' in the Mariana Trench of the Pacific, at around 10,916 metres (more than six and three-quarter miles) beneath the waves. That's deep enough to fit Mount Everest above it and still have more than two kilometres (nearly a mile and a quarter) to spare before reaching the surface.

Not all of the ocean is that deep, of course. To grasp that reality about our ocean planet, we can draw a 'hypsographic curve', which shows the percentages of the Earth's surface at different depths and altitudes relative to sea level. From that, we can see that the dizzying trenches, where the sea floor drops to more than six kilometres deep, are the exception rather than the rule. But shallow seas that fringe the land – the 'continental shelves' that are usually less than 200 metres deep – don't add up to much globally either. The average depth of the oceans is around 3.4 kilometres, and half of our planet is covered by water more than

3.2 kilometres (two miles) deep. Because the sun's rays only penetrate the top kilometre of the clearest seawater before they are quenched by it, the blue appearance of our world from space is just the reflection from that thin veneer of surface ocean. Most of the solid surface of our planet is ocean floor cloaked in darkness, beyond the reach of sunlight. So our 'blue marble' is really a dark, deep-ocean world.

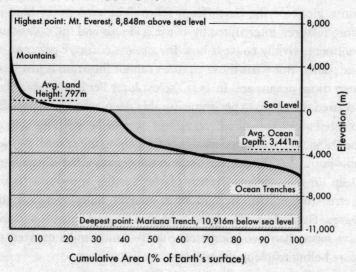

Hypsographic Curve

Highest point: Mt. Everest, 8,848m above sea level — 8,000

Mountains   4,000

Avg. Land Height: 797m   Sea Level   0

Avg. Ocean Depth: 3,441m   -4,000

Ocean Trenches   -8,000

Deepest point: Mariana Trench, 10,916m below sea level   -11,000

Elevation (m)

0   10   20   30   40   50   60   70   80   90   100

Cumulative Area (% of Earth's surface)

If we look at events such as the first human colonisation of Australia, we can see that our ancestors began crossing the oceans more than 40,000 years ago. But for almost all of history, what lies deep beneath their surface has remained unknown. We developed the technology to chart the ocean depths less than two centuries ago, and vehicles to withstand the crushing pressures of the abyss less than a century ago. That makes the twenty-first century a great time to be an ocean explorer, because the hidden face of our planet is now within our reach, and starting to give up its secrets.

*   *   *

Right now we have more detailed maps of the surfaces of the Moon and Mars than we do of the deep ocean floor. Because the surfaces of our celestial neighbours aren't covered by seawater, spacecraft can use radar to map them in detail from orbit. But seawater blocks the radio waves used by radar, so we can't bounce those signals off the ocean floor to map its terrain in the same way. With no seawater in the way, NASA's *Mars Odyssey* spacecraft has mapped the entire Martian surface to around a 100-metre level of detail, revealing any features such as craters or canyons that are larger than that size. At least 60 per cent of the Red Planet has also been mapped to a 20-metre level of detail by the European Space Agency's Mars Express mission, and China's *Chang'e 2* orbiter has mapped the surface of the Moon to a seven-metre level of detail. Meanwhile, the total area of the ocean floor that has been mapped so far to a 100-metre level of detail, using sonar from ships instead of radar from satellites, is about 15 per cent.

But having a more detailed map doesn't mean that we 'know more about' the Moon and Mars. The total amount of rock that has been analysed from the Moon to understand its geology is just over half a tonne, and the amount of rock analysed by rovers on Mars – and from very rare chunks of Mars that have tumbled to Earth as meteorites – is even less. Many thousands of times more samples and huge volumes of other data have been collected and analysed from the deep ocean to understand what's going on down there, from the processes that shape the ocean floor to the environmental conditions and patterns of life in deep waters. So despite lagging behind in the detail of our maps, we really know far more about the deep ocean than we do about other planetary bodies in the solar system. There's still a great deal to explore and discover in the oceans, of course – but what we now know about their depths is as astounding as the unknown that remains.

There are two fundamental steps in exploring the deep ocean, and the first is to make a map of whatever you want to understand. That might be a map of the features of the seabed in a particular area: for example, if you want to understand the geological processes that shape the ocean floor. But it could also be a map of anything else, such as the temperature of deep waters across a region, if that's what you're interested in exploring. And it could be a map of how something changes over time, so perhaps the more accurate term for this is really the 'mapping and monitoring' step.

Traditionally, this step involves heading out to sea with a research ship and using its equipment to measure whatever you're interested in for your map, such as the depth of the ocean in different locations. But satellites can also help us to measure some things, even if they can't peer directly to the bottom of the sea from space. And we now have 'autonomous vehicles' to help us, which are not connected to a ship or remote controlled at all. We're starting to be able to send those autonomous vehicles out instead of ships, to make measurements along pre-programmed survey routes, or even just drift around in the ocean collecting data.

The maps that we make show us where there are things that we don't yet understand, such as a rift valley in the sea floor that no one can explain, or a mysterious patch of slightly warmer water in the otherwise chilly depths. So the 'make a map' stage really detects anomalies, where there are things happening that challenge our current understanding of how the oceans work.

The second step in ocean exploration is to 'investigate the anomalies' revealed by our maps. Sometimes we might be able to investigate anomalies using the measurements that make up our new map: for example we can use them to improve computer simulations of how the ocean works. But typically the 'investigate the anomalies' step requires getting out there to study the anomalies in more detail, often close-up. That may involve using equipment aboard a research ship to collect samples or other data from the depths, or sending down deep-diving vehicles to work at the ocean floor.

We now have three types of underwater vehicles for working in the deep ocean, far beyond the reach of scuba-divers. First there are 'Human-Occupied Vehicles', which have the acronym 'HOVs': these are mini-submarines that can carry people into the deep ocean. In an HOV, we are not attached to the ship but are free to explore the depths on our own. During a dive we can decide exactly where to make measurements on the basis of what we are seeing, or exactly what to pick up with the craft's mechanical arms, to test our ideas about what's going on. The strong hull of the craft protects us from the pressure of the ocean outside, so we stay at normal surface conditions inside and avoid the complications of having to decompress like deep scuba-divers.

But building a craft that protects people from the crushing pressures of the abyss is expensive. So we also have 'Remotely Operated Vehicles', known as ROVs, which we control from a ship above via a cable, allowing us to send our minds, if not our bodies, into the deep ocean. ROVs transmit live video via their cable to a control centre aboard the ship, so their cameras become our eyes, and we can operate their manipulator arms to make measurements and collect samples just as if we were down there in a mini-submarine.

Human-Occupied and Remotely Operated Vehicles are both human-directed vehicles, and because they are under our direct control, they are great for the 'investigate the anomalies' step of ocean exploration. During this step, we test the ideas that we have about what might be going on – the possible explanations for the anomalies. For this, we typically need to collect very specific samples or data, and often develop our ideas further as we go along. And sometimes the missing puzzle piece that we need may be something unexpected that we come across by accident during our investigation. So we need to be directly involved, making decisions on the basis of what we're finding; we can't yet make a robot explorer that has the curiosity and creativity to replace us completely in making sense of the unknown.

At a 1956 meeting of US scientists to discuss technology to explore the deep, Allyn Vine of Woods Hole Oceanographic Institution, who subsequently led the creation of one of the most famous Human-Occupied Vehicles for science, said: 'I believe firmly that a good instrument can measure almost anything better than a person can if you know what you want to measure . . . But people are so versatile, they can sense things to be done and can investigate problems. I find it difficult to imagine what kind of instrument should have been put on the *Beagle* instead of Charles Darwin.'

As a third type of underwater vehicle, we have 'Autonomous Underwater Vehicles', which have the acronym 'AUVs'. These are not attached to a ship like a ROV, and not under our direct control; instead, we send them out on pre-programmed missions to collect data for us, which is great for the 'make a map' step and can sometimes help with the 'investigate the anomalies' step as well. We can leave them unattended to undertake their missions, freeing up research ships to carry out tasks with other equipment and enabling us to use our 'human-directed' technology more efficiently. Some AUVs can even be launched from shore, for missions lasting a month or more, swimming out to their survey areas without any ship required. One of their main limitations is battery power: unlike ROVs that are connected to a ship and draw power from it, AUVs typically carry their power with them in batteries, and their sensors and instruments use power every time they make a measurement. So for very long missions underwater, they can only use low-power sensors, which are adequate, for example, for measuring the temperature and salinity of the ocean; but if we wanted an AUV to use lights and cameras to film the ocean floor, this would deplete its batteries much more quickly, shortening the range of its missions.

Together, these three types of vehicles allow us to work routinely in the deep, and they are being used every day by research ships around the world as we gradually discover more about our oceans. But technology alone isn't enough to explore the ocean: it also needs diverse teams of people. For example, there are engineers, who design, build and maintain the equipment that we use, and our research ships don't go anywhere without a crew of professional mariners to run them, along with an organisation ashore to manage their logistics. We need scientists with different specialisms to interpret measurements and analyse samples, and computer experts to handle the vast volumes of different types of data now typically generated by an ocean expedition. Working together with such diverse teams – and helping them to work together effectively, if you are leading an expedition – is also one of the best parts of being an ocean explorer, right up there with seeing things that no one has ever seen before.

Today there is no longer a lack of technological capability holding us back in exploring the oceans: we can reach the greatest depths, if we can find the will to go there. And for reasons to explore the deep ocean, there's perhaps no finer summary than the words of John Steinbeck, in his 1966 letter to *Popular Science* magazine calling for the creation of an 'undersea NASA':

> *There is something for everyone in the sea – incredible beauty for the artist, the excitement and danger of exploration for the brave and restless, an open door for the ingenuity and inquisitiveness of the clever, a new world for the bored, food for the hungry, and incalculable material wealth for the acquisitive – and all of these in addition to the pure clean wonder of increasing knowledge.*[1]

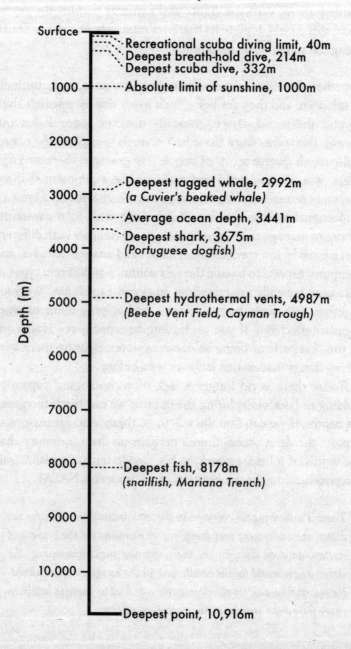

Ocean Depth Scale

Surface —

- Recreational scuba diving limit, 40m
- Deepest breath-hold dive, 214m
- Deepest scuba dive, 332m

1000 — Absolute limit of sunshine, 1000m

2000 —

3000 —
- Deepest tagged whale, 2992m
  (a Cuvier's beaked whale)
- Average ocean depth, 3441m
- Deepest shark, 3675m
  (Portuguese dogfish)

4000 —

5000 — Deepest hydrothermal vents, 4987m
  (Beebe Vent Field, Cayman Trough)

6000 —

7000 —

8000 — Deepest fish, 8178m
  (snailfish, Mariana Trench)

9000 —

10,000 —

Deepest point, 10,916m

Depth (m)

2

How do we map the oceans?

We can think of maps of coastlines as being the first ocean maps; they at least show the boundaries of the seas, even if they don't show features such as ocean currents or the depths of the ocean floor. Hecataeus of Miletus produced a map of the world that was known to the ancient Greeks, two and a half thousand years ago, improving on an even earlier but since-lost map by Anaximander. The Mediterranean of Hecataeus' map is recognisable to us in the shape of its coastlines and the locations of major islands. Fast-forward to the sixteenth century and we have beautiful works such as Abraham Ortelius's world map of 1564. On these maps the outlines of many of the continents look familiar, but the oceans themselves are blank in terms of any features such as currents or the undulations of the deep sea floor.

To start making a map of anything in the ocean, first you need to know where you are, which can be tricky for a ship out of sight of the shore, surrounded by nothing but the shifting skin of the sea. So the first maps of the ocean floor had to wait for some improvements in the technology for determining a ship's position at sea. Describing your position on the surface of the globe involves two coordinates: latitude, which measures degrees north or south of the equator, and longitude, which measures degrees east or west of a line running from pole to pole. Today we can get those two coordinates instantly from a Global Positioning System device that receives signals from satellites in orbit, but it used to take rather more effort.

Of the two coordinates, latitude is traditionally the easier one to measure – we can do it by, for example, the angle of the sun

in the sky at noon, or the angle of the North Star in the sky at
night if you are in the northern hemisphere – and mariners have
used and refined those techniques over centuries. But longitude
used to be more tricky to measure accurately. The principle
sounds easy enough: keep a clock running with the time where
you set out from, and note the time when the sun is highest in
the sky wherever you are. The difference between that 'local
noon' and twelve o'clock midday where you started will tell you
how far east or west you have travelled. The problem was having
a clock that keeps time accurately for months at sea, in damp
conditions aboard a rocking vessel. An error of a few minutes
could mean an error of several kilometres when estimating a
position, which could make the difference, in the middle of a
very large ocean, between arriving within sight of a tiny island
and missing it, or finding a safe passage rather than being
wrecked on rocky shoals.

In 1714, the British government established a competition to
develop better technology for determining longitude, and. John
Harrison was eventually awarded prizes for his increasingly accurate marine chronometers. Captain James Cook took a copy of
one of Harrison's chronometers on his second voyage exploring
the Pacific and Southern Ocean from 1772 to 1775, and he praised
its performance. Such equipment by the start of the nineteenth
century had become standard for ship navigation, allowing the
first 'oceanographers' to start making more accurate maps of the
ocean and its features using measurements taken in different
locations.

One of the first maps of a feature in the ocean was published
by Benjamin Franklin in the early 1770s, and it shows the Gulf
Stream – a strong current of warmer surface water that sweeps
north and east from Florida to Nova Scotia in the Atlantic. As
deputy postmaster general of British North America, and later
the first Postmaster General of the United States, Franklin
became aware that supposedly fast 'packet' ships bringing mail

from the south-western tip of England to New York took two weeks longer to arrive than merchant ships travelling from London to Rhode Island, even though the latter journey involved a greater distance.

Franklin consulted his cousin Timothy Folger, a merchant ship captain who had also sailed as a Nantucket whaler. Folger suggested that British captains of mail packet ships were not aware of the Gulf Stream and sailed against its current, which slowed them down. But American sailors knew about the current, thanks to the experience of their whalers in the area, and they kept outside it, which allowed them to make quicker westward journeys. Folger even recalled hailing westbound British mail ships in the Gulf Stream, to advise them that they would make better progress if they moved out of the current, but he reported that 'they were too wise to be counselled by simple American fisherman'.[2] Franklin asked Folger to sketch the boundaries of the Gulf Stream, and then had copies of the map printed for British mail ship captains, who unfortunately still chose to ignore it. But the map, which Franklin later reproduced in the *Transactions of the American Philosophical Society* in 1786, is remarkably accurate in its depiction of the boundaries of the Gulf Stream; we can see the same boundaries in satellite measurements of ocean surface temperatures today.

With more reliable technology for knowing their location, the early oceanographers of the nineteenth century started mapping the contours of the sea floor far below. To measure the depth of the ocean, mariners dropped weighted lines from their ships, judging when the weight hit the seabed from changes in the rate at which the line unreeled, and recording the length of line paid out as the depth at that location. On 3 January 1840, for example, James Clark Ross on his way to the Antarctic used this technique to measure an ocean floor depth of more than four kilometres, setting a record for the time.

The first maps showing the hidden landscape of the ocean floor therefore began to emerge in the mid-nineteenth century, and included those made by Matthew Fontaine Maury of the US Navy. Maury is one of my heroes from the early history of ocean exploration; he was born in 1806 and his older brother John was a distinguished young officer in the US Navy. John died of yellow fever at sea when he was twenty-eight, and Matthew followed his brother into the navy, despite the objections of their grieving father. Matthew served aboard the USS *Vincennes* during the first round-the-world voyage undertaken by the US Navy, but he broke his leg in a stagecoach accident at the age of thirty-three, which put an end to his seagoing career. By then he had already published two books on navigation, so he pursued the study of navigation and meteorology ashore, becoming superintendent of the navy's Depot of Charts and Instruments, which later became the United States Naval Observatory.

In his role at the Naval Observatory, Maury compiled charts of wind and currents in the north Atlantic, and helped to standardise the methods for making oceanographic measurements at sea. In 1853 he published the first sea-floor map of the north Atlantic, sketching the contours of its unseen terrain from the sparse dots of depths measured by US Navy ships. The first map required rapid updating, as each depth measurement explored such a vast unknown. So one year later, Maury published a revised version of the map that included further depth measurements made in the middle of the ocean by the USS *Dolphin*. Those additional measurements revealed a seabed plateau north of the Azores, slightly shallower than the deep ocean basins on either side of it. The plateau was actually the first glimpse of the Mid-Atlantic Ridge, which forms part of our planet's greatest geological feature, a vast undersea volcanic rift known as the mid-ocean ridge. The plateau on Maury's map of 1854 looked like a good place to attempt to lay the first undersea telegraph cable between Europe and America, and so the first major feature discovered from mapping the ocean floor was nicknamed 'Telegraph Plateau'.

Let's just pause at this point to take stock: although our ancestors perhaps started crossing the oceans at least 40,000 years ago, our knowledge of the deep ocean floor really begins here, less than two centuries ago. To put that another way: if our 40,000-year history of crossing the oceans is represented by a 24-hour clock, with midnight as the present day, then we started to fathom the ocean depths just after 11:52 p.m., less than eight minutes ago.

Dropping a weighted line to measure the depth of the ocean can be cumbersome, requiring a ship to stop for several hours – often with much sweaty hauling of kilometres of rope by the sailors – just to get one depth measurement in one location. The round-the-world voyage of ocean exploration made by HMS *Challenger* from December 1872 to May 1876 made several hundred depth measurements using that technique – a great achievement for the time, but still only a few pinpricks in the vast unknown of the ocean floor. The nineteenth-century polymath William Thomson, who later became Lord Kelvin, devised a new depth-measuring device using piano wire to plumb the ocean depths, which proved quicker and more reliable than using rope. We may therefore be able to thank John Isaac Hawkins, born in Taunton in Somerset in 1772, for an unsung yet musical contribution to deep-sea exploration. In 1826, Hawkins invented an upright piano whose popular ownership among the growing middle classes of the nineteenth century led to the mass production of piano wire, making it cheaply available in plentiful quantities for mapping the ocean depths.

Thomson's piano-wire device was used successfully by the USS *Tuscarora* under the command of George Belknap to explore the depths of the Pacific in 1873 to 1874, scouting for a possible route for a sea-floor telegraph cable between the US and Japan.

Using Thomson's device, the *Tuscarora* measured a depth of 8,513 metres in what we now call the Kuril–Kamchatka Trench, which was greater than the depth measured by HMS *Challenger* in the Mariana Trench in 1875. The voyage of the *Tuscarora* also revealed the existence of underwater mountains known as seamounts, as described in Henry Cummings's synopsis of the expedition from October 1873:

> While sounding on the afternoon of the 27th, some 140 miles off the coast of California, and expecting a depth of 1,600 or 1,700 fathoms (the previous cast having been 1,689 fathoms), the lead suddenly brought up at a depth of 996 fathoms . . . We then sounded round this locality and found that a rocky submarine peak, 4,000 feet in height, existed in this part of the ocean.[3]

Charles Sigsbee of the US Navy refined Thomson's depth-measuring device further during the expeditions of the schooner *George S. Blake* in the Gulf of Mexico from 1875 onwards. His highly efficient 'Sigsbee Sounding Machine' could be powered by a small steam engine, and it set the standard for depth-measuring technology for several decades afterwards. Sigsbee's machine helped the *Blake* to make around three thousand depth measurements in the Gulf of Mexico – nearly ten times the number made worldwide by HMS *Challenger* earlier in the same decade. Having so many measurements reduced the guesswork in 'joining the dots' to sketch the contours of the ocean floor, and the sea-floor chart of the Gulf of Mexico produced from the *Blake*'s survey was the first really reliable sea-floor map of a deep ocean basin.

'Sounding' the ocean describes measuring its depth by any technique. It comes from the Norse word *sund*, meaning gap, hence its use in coastal names that indicate a channel for ships, such as the inlets of Plymouth Sound in the southwest British Isles and Puget Sound in Washington State. But the invention of echosounding in the early twentieth century enabled sound – in the sense of a noise – to be used for measuring the depth of the ocean floor, instead of a weighted line. The development of echosounding meant that ships could measure the depth of the sea floor beneath them almost continuously, instead of having to stop a ship for several hours to lower a weighted line to get a single measurement in just one location.

The principle behind echosounding is to send a pulse of sound from a ship to the seabed below, and measure the time it takes for an echo to return to the ship. If we know the speed of sound in seawater, then the time interval between the outgoing pulse and its returning echo can be converted into the distance travelled by the sound. You can try echosounding on land, near a cliff or perhaps in a large courtyard, by clapping your hands loudly and timing the echo as the sound bounces back to you from the cliff or courtyard wall. The speed of sound through air is much less than its speed through seawater, because air is less dense: sound travels about 340 metres per second through air, compared with nearly 1,500 metres per second through seawater. So if your cliff or courtyard wall is 100 metres away, you should hear the echo about two-thirds of a second after your clap – the time it takes for the sound to make the 200-metre round trip.

The idea of using 'echosounding' to measure the depth of the ocean floor was actually described by Maury in 1855:

> By exploding heavy charges of powder in the deep sea, when the winds were hushed and all was still, the echo or reverberation from the bottom might, it was held, be heard, and the depth determined from the rate at which sound travels through water.[4]

Maury reported that 'no answer was received from the bottom', however, when this method had been attempted. But rather than using gunpowder and relying on human ears, modern echosounding uses electronics to produce clearer 'pings' of sound and detect their echoes. With echosounders, oceanographers could trace detailed undulations of the seabed wherever their ships travelled. From 1925 to 1927, the expedition of the German ship *Meteor* used an echosounder to record 67,000 depth measurements as it criss-crossed the Atlantic thirteen times – a huge advance on the few hundred soundings made worldwide by HMS *Challenger* five decades earlier. The voyage of the *Meteor* revealed that the Mid-Atlantic Ridge runs from north of the Azores into the southern hemisphere, eventually turning eastwards towards the Indian Ocean. So the 'Telegraph Plateau' that first appeared on Maury's map of 1854 became part of something much greater.

The first global map of the ocean floor made using data from echosounders was a collaboration by scientists Marie Tharp and Bruce Heezen with artist Heinrich Berann. Marie Tharp is another of my heroes of ocean exploration; in the 1950s she began collating the information coming ashore from expeditions to build up maps of the ocean floor in different regions, culminating in the global 'Floor Of The Ocean Map' published in 1977. The 'Floor Of The Ocean Map' was based on hundreds of survey lines from ships around the world, with gaps between those actual measurements filled in by informed geological guesswork from Tharp and Heezen, and artistic interpretation by Berann. I think the 'Floor Of The Ocean Map' is as inspirational as the first 'blue marble' photograph of the Earth from space, because it sketched the hidden face of our planet for the first time. The map has been reproduced several times in *National Geographic* magazine, and as a child I spent many happy hours poring over it, wondering about the mysterious-sounding places that it showed, from the Reykjanes Ridge to the Bouvet Triple Junction, none of which were mentioned in my geography classes at school.

Modern research ships still map the ocean with sound, but the echosounders of the early twentieth century have now been refined to produce an even better view of the sea floor. Rather than sending out just one 'ping' at a time to measure the depth immediately under the ship, today's systems send out a burst of 'pings', which spread out like a fan and simultaneously map a much broader strip of seabed as the vessel steams ahead. These 'multibeam' sonar systems therefore map far more of the sea floor at once than old single-beam systems, and in much greater detail, typically revealing features around 100 metres in size on the ocean floor thousands of metres below. To ensure accuracy, hydrographers also lower probes during surveys to measure how the density of seawater beneath the ship varies, from which they can calculate the exact speed of sound through it, to turn the timings of the echoes into reliable depth measurements.

The angle at which the sonar beams fan out from the ship means that the strip of seabed being mapped is several times as wide as the depth of water – so if you are mapping in an area around one kilometre deep, the multibeam sonar can usually 'see' a strip of seabed several kilometres across, stretching out to either side of the ship. Planning parallel survey lines for the ship so that the strips overlap along their edges – a pattern known as 'mowing the lawn' – then builds up a complete map of the ocean floor over the survey area.

In 2009, there was a flurry of misguided media excitement when someone browsing the online maps in Google Earth spotted a grid-like pattern on the ocean floor in the Atlantic. It looked like the regular arrangement of a city's streets, leading to headlines about the possible discovery of the sunken city of Atlantis. But it was really the 'mowing the lawn' pattern of a survey ship that was collecting better ocean-depth data for the map in that area, which stood out against the background of less detailed measurements.

Even when a research ship is just travelling from one location to another, rather than 'mowing the lawn' to map an area, we usually still map a strip of seabed along our track, helping to build up a more detailed map of the global ocean floor day by day as research ships travel the seas. 'Watchkeeping' duty while the journey is underway – which involves keeping an eye on the depth measurements and occasionally tweaking the instruments as conditions change – is one of my favourite jobs at sea. There's a special thrill in watching a previously unknown underwater mountain gradually appear on the mapping screen as your ship passes over it, and realising that you are the first to see it. And there are plenty of sea-floor features still out there waiting to be discovered: in 2009 my colleagues and I were working near the South Sandwich Islands in the Antarctic when our ship's sonar revealed a sea-floor crater 4 kilometres in diameter – about one-quarter the size of Crater Lake in Oregon – and 1.6 kilometres deep that no one had previously known was there.

If we want to map the smaller features of the sea floor, we need to take our sonar much closer to it. To do that, we can put multi-beam sonar systems on underwater vehicles, and either tow them on cables or programme them to 'fly' close to the ocean floor. The area that they can map becomes much smaller because the 'fan' of sonar beams now begins much closer to the seabed. The size of the features that they can detect becomes much smaller too; flying a multibeam sonar system about 50 metres above the seabed typically allows us to map a strip just a couple of hundred metres wide, and detect individual boulders within that strip of just a couple of metres in size – or potentially find shipwrecks or debris from crashed aircraft. In comparison, from 50 metres above the sea floor we can't see it through the water with cameras, and if we go closer to the sea floor so that it comes into view, the area we can see becomes much smaller than a sonar strip. So sonar is our primary tool for mapping and searching in the deep ocean; visual inspection is usually then part of the subsequent 'investigate the anomalies' step in ocean exploration.

Meanwhile, although seawater blocks radar signals from reaching the ocean floor from space, satellites can use radar to measure bumps and dips in the surface of the sea very accurately. It may seem incredible – and it involves a lot of maths – but tiny undulations in the surface of that liquid skin can now also give us a glimpse of the landscape of the ocean floor far below. Where there is a large underwater mountain poking up from the surrounding sea floor, for example, a minuscule local increase in gravity resulting from its mass pulls seawater into a slight bump above it. If on the other hand there is an ocean trench, the weaker local gravity resulting from the more distant sea floor produces a comparative dip in the ocean surface. Reading these subtle bumps and dips in the sea surface from space is an astounding feat, requiring measurements over many years to subtract the effects of wind-blown waves and tides, and using lasers to track the flight paths of the satellites very precisely. But it is possible to use this technique to map the large-scale features of the ocean floor, such as the entire mid-ocean ridge, the vast abyssal plains and all the ocean trenches that cut across our planet's surface like scars.

In 1997, David Smith from Scripps Institution of Oceanography in San Diego and Walter Sandwell from the US National Oceanic and Atmospheric Administration published the first 'gravity-derived' map of the entire ocean floor made from satellite data, which showed sea-floor features as small as twenty kilometres across. And in 2014, having crunched through even more data from newer satellites, Smith and his colleagues published an updated global map, this time revealing sea-floor features as small as five kilometres in size. So while taking our sonar systems closer to the ocean floor gives us increasingly detailed maps of increasingly smaller areas of it, some nifty tricks with satellites have now given us large-scale maps of the entire ocean floor.

We sometimes hear that the oceans are 'only five per cent explored' or 'ninety-five per cent unexplored', but that's a popular myth rather than really accurate, because it depends on what we mean by 'explored'. If 'explored' means 'mapped', then we now have a map of 100 per cent of the ocean floor, albeit only at a five-kilometre level of detail. Some people question whether that's really a map of the ocean floor, because it's a cunning estimate of undersea terrain from satellite measurements of the sea surface, rather than being based on more direct measurements of depth using sonar. But I would reply: can you use the satellite-based map of the ocean floor to find your way to the large underwater features shown on it? The answer to that is 'yes', so it is functionally a map of the entire ocean floor. That therefore casts some doubt on the 'only five per cent explored' claim, though of course the satellite-based map doesn't reveal everything that is down there, as it only has a five-kilometre level of detail.

The total area of ocean floor that has now been mapped to around a 100-metre level of detail by multibeam sonar systems from ships is probably around 15 per cent, which is an area roughly the same size as the continents of Africa and North America combined. It's difficult to put a precise percentage on it, because the data are spread across various archives, and the area mapped is growing all the time. But the 100-metre level of detail is often what people have in mind when they talk about 'mapping' the ocean floor, and it may be the source of the 'only five per cent explored' myth, because although that figure is certainly out of date now, a couple of decades ago the coverage *was* only about 5 per cent.

The fact that the majority of the ocean floor remains unmapped by sonar from ships is not the result of a lack of technological capability – it is just a consequence of the size of the oceans. A modern research ship could in theory complete the sonar mapping of the entire ocean floor, but it would take several centuries to do so, even if it was dedicated to that task all the time. Scientists in different countries are therefore coordinating their efforts and ships to try to reach that goal sooner, and there is also a new international XPRIZE competition to develop quicker methods for mapping the ocean floor in detail, for example by using squadrons of autonomous vehicles.

Meanwhile, the area of ocean floor that has been mapped to the highest level of detail, revealing objects a couple of metres across by using sonar from underwater vehicles close to the seabed, is probably only one-twentieth of 1 per cent, roughly equivalent in size to the state of Oklahoma in the US, or slightly larger than the area of England and Wales combined. And the area of ocean floor that has actually been seen by human eyes peering through the portholes of deep-diving mini-submarines, or via towed cameras and Remotely Operated Vehicles, is even smaller still – probably ten times less, which would be an area slightly smaller than Wales, or about twice the size of the state of New Hampshire.

So on one hand, the 'ninety-five per cent unexplored' myth doesn't do justice to the large-scale map that we have of the entire ocean floor from satellites. But at the other extreme, if we take 'explored' to mean 'seen by human eyes', then much less than 5 per cent of the ocean floor has been 'explored', and that's also ignoring the enormous volume of water that forms the interior of the ocean above the ocean floor. Furthermore, does glimpsing a place just once mean it has been 'explored'? The local woods where I walk my dog look very different in winter compared with summer, with different species flourishing in different seasons. Should I have considered them 'explored' after my first visit in just one season? The 'ninety-five per cent unexplored' or 'only five per cent explored' meme is a popular rallying cry for ocean exploration – and I've been guilty of using it myself in the past – but it doesn't really reflect what we know and don't yet know. Exploring our world starts with making a map, but arguably it doesn't have an end.

3

Are there really 'sea monsters' in the ocean?

Medieval map-makers often doodled sea monsters in the otherwise blank ocean spaces on their maps. They were generally inspired by tales from mariners and used to illustrate or represent the dangers of the deep. But at least some of their artistic creations were based on encounters with real animals, so the oceans do contain some creatures that match the myths. Large marine mammals, in particular, made an impression on early seafarers, with many illustrations on early maps showing creatures that spout and snort like whales; sometimes they're even shown capsizing ships. A male narwhal with its single long 'tusk' – actually a very long canine tooth – may have been the inspiration for the 'sea unicorn' on the Carta Marina of 1539 by Olaus Magnus. And these encounters also sparked the imagination as recorded in ships' logs as well as on maps. On 9 January 1493, Christopher Columbus recorded seeing three 'mermaids' in the Caribbean, noting that they were 'not as beautiful as they are painted' – perhaps not surprising, as they were most likely manatees.

Some legends of sea serpents may have come from rare encounters with giant oarfish – the longest bony fish in the ocean, with a ribbon-like body that can exceed ten metres. They roam at depths down to 1,000 metres, and although seldom found at the surface, they do occasionally wash up on shores. But more of these tales of sea serpents may have stemmed from sightings of large squid such as *Architeuthis dux* – the Giant Squid. There have been reported sightings of these creatures since at least the first century CE; Pliny the Elder in his *Naturalis Historia* gives a second-hand account of an enormous squid-like animal raiding fish-pickling ponds on Spain's Atlantic coast, although his account describes the animal climbing a tree to get over palisades around the ponds – although octopuses sometimes crawl out of the sea, squid are not known to do that.

The Giant Squid's invertebrate body, so different in appearance from that of the more familiar whales or fish, may have led to confusion and been mistaken for a serpent's body when dead ones washed up on beaches. Its eight arms and two feeding tentacles, which can reach ten metres long when fully extended, may also have been mistaken for serpents when mariners came across specimens at the surface. The discovery of an eight-metre tentacle in the stomach of a sperm whale in 1783 certainly caught the imagination of French naturalist Denys de Montfort, who became fascinated with the possibility of giant octopuses attacking ships; his depiction of such a creature rising from the sea to entwine its arms around the masts of a ship is an iconic piece of sea monster art. They have always been elusive, and the first photographs of a Giant Squid alive in the deep ocean were taken in 2004 by Japanese scientists who lowered some bait with a camera to a depth of 900 metres, and the first video of one in the deep sea was only taken in 2012. There is another species of very large squid that has not yet been filmed alive in the deep, although specimens are sometimes caught by deep-water trawlers or seen near the surface: at around fourteen metres long to the tips of its tentacles, the Colossal Squid (*Mesonychoteuthis hamiltoni*), which lives in the seas around Antarctica, is about the same length as a Giant Squid, but heavier in build. And other large squid lurk in the deep, such as the big-fin squid group of species (*Magnapinna*), which have exceptionally long arms compared with their bodies. Big-fin squid have occasionally been spotted from Human-Occupied Vehicles and Remotely Operated Vehicles since 1988, usually at depths greater than two kilometres.

John Steinbeck wrote that 'an ocean without unnamed monsters would be like sleep without dreams', and our fascination with what might dwell unknown in the deeps remains strong. Video clips of strange creatures glimpsed from underwater vehicles often go viral online, such as species of so-called 'trash bag' jellyfish (*Deepstaria*) – this creature's billowing translucent body, more than a metre across, is patterned with a mysterious-looking hexagon-mesh of lines that are actually part of its digestive system. And if the carcass of an uncertain large animal rolls ashore onto a beach, it typically triggers speculation in the media, even after marine biologists have managed to identify it. These 'beach blobs' almost always turn out to be the decomposed remains of whales, rather than specimens of some new sea monster. For example, by analysing fresh and museum-archived samples, Sidney Pierce of the University of Southern Florida and colleagues have shown that the thirteen-tonne 'Chilean blob' of 2003, the Nantucket blob of 1996, two Bermuda blobs from the 1990s, the Tasmanian west coast monster of 1960 and the 'giant octopus of Saint Augustine' from 1896 were all washed-up whales.

The largest animal that has ever lived does roam our oceans today: at around 125 tonnes, an adult blue whale's body mass is nearly twice that of the biggest 'titanosaur' dinosaurs, and equivalent to about two thousand people – the population of a village, perhaps. And other large animals are still being discovered in the oceans, such as the megamouth shark, which grows to around five metres long and was described as a new species by scientists in 1983, from a specimen found near the Hawaiian Islands in 1976. It uses its gaping 'megamouth' to filter plankton to eat, and lives as deep as 1,500 metres down in the open ocean. Recent discoveries like this raise the question: how many undiscovered large animals could still be out there in the oceans? By looking at the rate at which new large species are still being discovered, we can actually make a guess at that, as a 'known unknown' in the deep.

*   *   *

Imagine you have – or better still, actually equip yourself with – a bag of sweets such as wine gums, with six different types of wine gum inside. The first time you put your hand into the bag without looking and take out a sweet, it will inevitably be a type that you haven't taken out of that particular bag before. The next time you dip your hand in and take one out, there's a good chance it will be one of the five other types that you haven't yet taken out of that bag. But there's also a small chance that it could be the same type that you have already taken from the bag. As you keep taking sweets out, building up a collection of the different types, the chance of you taking out a type that you haven't seen from that bag before becomes less and less. Eventually, once you have taken out a collection that includes all of the different types of sweet, then no matter how many more sweets you take out of the bag, you won't find any new types in there that you haven't already seen.

If we have already found all the large species in the ocean, then no matter how much more we look, we won't find any new ones – the rate at which we find new large species will be zero if we already know them all. And if we're close to finding all the large species in the ocean, then we won't find new ones very often as we continue to look. So if we plot the number of animal species more than two metres long that are known in the ocean against time, counting them from the descriptions that scientists publish when they present a new species to the world, we can see whether that 'discovery curve' has flattened off. The graph does look close to flattening off, but it hasn't quite yet; and by using some maths to extrapolate its curve, researchers can estimate that there are probably around ten more species of large marine animals still out there waiting to be discovered.

But just because we haven't fully explored the ocean doesn't mean 'anything is possible', and creatures from modern media myths such as the 'megalodon' supershark and mermaids are not waiting for us out there. Although a couple of TV shows and

Discovery Curve

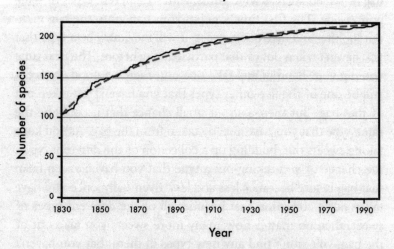

(data from Charles Paxton, 1998, "A cumulative species description curve for large open water marine animals", Journal of the Marine Biological Association of the UK, volume 78, pages 1389-1391)

films have pretended that megalodons and mermaids could be living in the oceans today, they are science fiction: in the case of the TV shows, their 'investigations' involved actors playing fictional scientists, with computer-generated images as 'evidence', and an easily missed disclaimer flashed up in the end credits. Whatever is still undiscovered has to be consistent with what we already know about how the oceans work, and unless we're wrong about everything in the picture we've been building up so far, megalodons and mermaids don't fit with it.

Otodus megalodon (also known as *Carcharocles megalodon*, while palaeobiologists figure out its family tree) was a real species of shark that hunted in the oceans from more than 20 million years ago until around 2.6 million years ago. Its teeth, which can be eighteen centimetres long, are common around the world as fossils. Complete cartilage skeletons of sharks do not fossilise well, so biologists have had to estimate from its teeth and rarer fossil vertebrae that the average length of an adult megalodon would probably have been just over ten metres, which is nearly as long as a bus. Some of them may even have reached eighteen metres, which is about three times the length of the largest reliably measured Great White Shark today. Megalodons preyed on marine mammals, and particularly smaller species of whales that lived at the same time, whose fossil bones show bite marks that match megalodon teeth.

At the time that megalodons disappear from the fossil record, the oceans were changing, becoming colder and dropping in sea level as more water was locked up in ice at the poles. Ocean ecosystems changed too at that time, with several of the smaller whale species that megalodons preyed on going extinct. If megalodons were somehow still alive in the oceans today, we would be finding their modern teeth as well as fossil ones, because shark teeth drop off and are replaced throughout their lives. And as huge top predators, they would need to eat a lot of large prey to survive, and therefore have to hunt over a very large area, so we should have come across signs of them – they cannot survive cooped up in a cave somewhere like the 'living fossil' coelacanth fish. Contrary to the fun idea in sci-fi novels and the film *The Meg*, it couldn't be lurking at the bottom of an ocean trench either, because the chemistry inside the bodies of all species of sharks prevents them from living that deep.

As for mermaids: there is no biological, archaeological or fossil evidence that aquatic humanoids exist or have ever existed. If they did exist, there would be a fossil trail of their evolution, and there should be related species sharing some of their characteristics, like the other members of the primate group to which we belong. There is a very rare human condition called sirenomelia, also known as mermaid syndrome, in which the legs of a developing foetus are fused together. Cases of this condition may have contributed to the legend of mermaids, along with sightings of marine mammals such as dugongs and manatees by seafarers who had been away from home for a long time.

4

What's the ocean floor really like?

Our increasingly detailed maps of the ocean floor show that its terrain is more varied than people used to think. I think that's perhaps one of the biggest revelations from ocean exploration in the late twentieth century: the deep sea floor isn't just a monotonous flat landscape of fine mud, or 'great grey level plains of ooze' as Rudyard Kipling described it in his poem 'The Deep-Sea Cables' of 1893. Instead, the ocean floor has as rich and varied a landscape as we find above the waves. So for an insight into what the ocean floor is really like, let's imagine going for a walk across the bottom of the Atlantic from Lisbon in Portugal to San Juan in Puerto Rico, following a slightly wandering route to take in several features.

If we start our journey within sight of the São Julião da Barra Fort that guards the entrance to the River Tagus near Lisbon, and head south-west away from the shore, then for the first twenty-five kilometres or so we cross a narrow strip of 'continental shelf' that is less than 150 metres deep here. The sea floor deepens very gradually across this shelf, with an overall downwards angle of less than half a degree, which we would barely notice if we were walking on it.

At around the 150-metre depth contour, the seabed starts to slope more steeply to greater depths, as we move from the continental shelf on to the 'continental slope'. Over the next forty kilometres of our hike out into the Atlantic, the sea floor drops to more than four kilometres deep. That's an overall downward angle of around ten degrees for this continental slope, and in some places the slope can be much steeper, with some cliffs and rocky outcrops. There are also a couple of underwater canyons carved into the slope here – the Lisboa and Setúbal Canyons – that would look quite spectacular if the water were removed so that we could look down into them. The canyons can act like funnels, with strong currents slipping down them and carrying sediments from the shallow continental shelf, which makes canyons an ideal home for types of marine life such as sponges and corals that filter a meal from those passing particles.

At the base of the continental slope, the sea floor flattens out. For the next 100 kilometres of our journey, the depth only increases by another 500 metres or so, as we head out into the middle of the Tagus Basin. That very gentle downwards angle of perhaps half a degree overall, similar to the continental shelf, would scarcely be noticeable underfoot.

Now we're at five kilometres deep, in the middle of a huge shallow-sloping bowl of soft mud sea floor about 400 kilometres across, which forms the 'abyssal plain' of the Tagus Basin. The sea floor of the abyssal plain is made of fine mud, formed by sediments falling perpetually from above and consisting of the dead bodies, excrement and detritus of everything living in the overlying water. Over millions of years those sinking sediments blanket the sea floor in a layer that can be a kilometre thick and proportionally as flat as a pancake at its surface, though in some areas currents can shape the sediments into ripples and even create features like dunes. Many of the inhabitants of the abyssal plains make their living from what falls from above, burrowing through the sediment or scraping across its surface to extract what nutrition they can from it. Others sniff out bigger 'food falls' where the dead bodies of larger animals, such as fish, jellyfish and even whales have sunk to the sea floor.

If we turn south at this point, then the peaks of two undersea mountains come into view, with a saddle-like ridge stretching forty-five kilometres between them – or at least they would be visible on a clear day if there wasn't seawater in the way. Those peaks rise five kilometres from the Tagus Basin to a depth of just fifty metres at their tops. The overall incline up their sides is about 1 in 6, which makes for quite a climb on foot, and there are some sheer cliffs and rocky outcrops. These two peaks and the saddle between them form Gorringe Bank, which was discovered in 1875 by the USS *Gettysburg* under the command of Captain Henry Honeychurch Gorringe, using Thomson's piano-wire depth-measuring device.

Underwater mountains are called seamounts if they rise more than one kilometre above the surrounding ocean floor, and there are more than 39,000 of them worldwide. Most seamounts are created by undersea volcanoes, but Gorringe Bank was formed where the seabed was pushed up by two of the plates of the Earth's crust sliding past each other. The slopes of seamounts are often too steep in places for sediments to settle on them, so seamounts provide extensive rocky habitats in the deep ocean, in contrast to the muddy areas of the abyssal plains. And ocean currents usually accelerate over seamounts, rather like the flow of a river surging as it is forced over and around rocks. A rocky sea floor on which to anchor, and faster currents bringing more food particles, therefore makes the tops of seamounts a good place for filter-feeding animals to grow, such as deep-sea corals and sponges. Fish and large ocean-wandering animals often also aggregate over shallower seamounts to feast on this bounty, so seamounts are often hotspots for marine life, and home to different species than the abyssal plains that they punctuate.

More than 800 species of marine life have been recorded here on Gorringe Bank, from the kelp forests at its peaks to the gardens of deep-water corals and sponges that crowd its flanks, but recent surveys have found discarded fishing gear and other litter here as well. Dropping back down to five kilometres deep on the south side of Gorringe Bank, we arrive at another small abyssal plain – the Ferradura Abyssal Plain. If we turn west here, then after 600 kilometres and a bit of weaving between a few other seamounts, we come to the northern end of the much larger Madeira Abyssal Plain. Swinging to the south-west for about 800 kilometres across its grey mud surface, we reach another spectacular underwater mountain on its far side.

The Great Meteor Seamount rises from more than four kilometres deep to a wide plateau around 150 to 300 metres deep at its summit, and it was discovered by the German ship *Meteor* in 1938. Its flat-top appearance is strikingly different from the two peaks and saddle of Gorringe Bank, and unlike Gorringe Bank this seamount grew as an undersea volcano. But it is similarly a hotspot for marine life; for example there are meadows of glass-skeleton deep-sea sponges on its sides at around 2.7 kilometres deep. It even has its own ocean circulation, with currents that swirl around its flat top and potentially corral some of its weaker-swimming inhabitants.

If we turn west and climb down from the Great Meteor Seamount back to four kilometres deep, the terrain starts to become more rugged. Our path to the west now takes us over 'abyssal hills', each rising and falling a couple of hundred metres over a few kilometres, which are actually our planet's most abundant surface feature. After about 800 kilometres of undulating abyssal hills, the overall terrain starts also to rise very gently ahead of us as we continue further west. We are now climbing the flanks of the Mid-Atlantic Ridge, which runs approximately north–south across our path. The Mid-Atlantic Ridge is part of our planet's hidden geological backbone, which is known as the mid-ocean ridge. We can trace the mid-ocean ridge all around the world from the Arctic through Iceland – where it briefly peeps above the waves – down through the Atlantic, into the Indian Ocean, down towards the Antarctic, across the South Pacific and along the western coasts of South and North America, and it also branches out in some other places.

The mid-ocean ridge forms where the huge plates of the ocean crust are being pulled apart by forces inside the Earth, and it's therefore a vast volcanic rift, where erupting lava creates new ocean floor. The crest of the ridge rises above the abyssal plains, and along our path here, the sea floor rises to around two kilometres deep over the next 500 kilometres of our journey towards it. The fine mud that blankets the sea floor gradually becomes more sparse on the way towards the ridge, exposing the bare volcanic rock of the ocean crust. At the crest of the ridge there is a rift valley, nearly a kilometre deep and about twenty-five kilometres across, blocking our way. The walls of the rift valley are steep – around 45 degrees overall, with vertical rock faces in some areas – so it would be a challenge to cross on foot.

The floor of the valley lies at a depth of three kilometres where we are crossing it, with some long mounds of freshly solidified lava, in shapes looking like dollops squeezed from a toothpaste tube, running along it. Elsewhere on the valley floor there are some flat areas where the solidified lava looks smoother, almost like the surface of a frozen lake, and in some places that smooth surface appears to have collapsed, revealing empty chambers below where lava has drained back deeper into the crust. There are even some bizarre pillars of basalt rock, left upstanding as columns that supported the now-collapsed roof of those chambers. Towards the valley walls there are also some scree-like jumbles of rock fragments, where solidified lava flows have been broken up by geological faulting.

At the top of one of the lava mounds running along the valley, there are plumes of smoky-looking water twisting up into the ocean above, which look like the steam plumes rising from geysers in Iceland. Those smoky-looking plumes of mineral-rich water come from undersea hot springs, known as hydrothermal vents. This particular cluster of them at 29 degrees north on the Mid-Atlantic Ridge is called the Broken Spur Vent Field, and these are the first hydrothermal vents that I investigated in my research, after they were discovered during an expedition of the British research ship RRS *Charles Darwin* in 1993. The volcanically heated water that gushes out of them builds slender mineral spires on the sea floor, reaching ten metres tall, which are festooned with deep-sea animals.

Having negotiated the rift valley to continue on our way, we follow a route to the west that runs down the gently sloping but rugged flanks of the other side of the Mid-Atlantic Ridge for another 500 kilometres, dropping back down to five kilometres deep. Continuing west for yet another 500 kilometres across yet more abyssal hills, and then turning south-west for 1,000 kilometres of similarly hilly terrain increasingly blanketed by sediments, we eventually reach the edge of the Nares Plain. This abyssal plain, similar to some of the other smaller ones we have encountered, is like a shallow-sloping bowl, about 600 kilometres across and reaching six kilometres deep at its centre.

At the southern edge of the Nares Plain, we're just 150 kilometres from our destination in Puerto Rico, but the ocean floor has one more surprise for us at the end of our journey. Here the seabed starts to get deeper, dropping to more than seven kilometres deep. This is the Puerto Rico Trench: the USS *Dolphin* measured a depth of seven kilometres here in 1852, and the USS *Milwaukee* sounded a depth of 8,740 metres in the trench in 1939. This is the deepest point in the Atlantic, since named the Milwaukee Deep. The French bathyscaph *Archimède* brought people more than eight kilometres deep here in 1964, and although people had previously gone deeper in the Mariana Trench of the Pacific, the dives here resulted in the first scientific paper to describe marine life from first-hand observations at the bottom of an ocean trench.

Ocean trenches form where one plate of the ocean crust collides head-on with another – the opposite of what happens at the mid-ocean ridge, where the plates are rifting apart. The slow-motion collision inevitably pushes one plate below the other, forcing it down into the Earth's mantle. These 'subduction zones' create the trenches, as the sea floor bends down on its way to be melted and ultimately recycled in the mantle below. Although few vehicles have explored the ocean trenches, scientists have been able to study them by lowering other equipment from ships – albeit always with patience, waiting many hours for the ship's winches to pay out and haul in the kilometres of wire needed to reach down into the 'hadal' environment of trenches, which gets its name from the Hades of Greek mythology. More recent insights have come from 'landers', which are not attached to a ship by a long wire but instead sink to the ocean floor on their own, armed with cameras, sensors and often bait to attract marine life. Scientists aboard the ship far above can then signal the lander to release its ballast weight and float back up to the surface with its cargo of data.

The Puerto Rico Trench is nearly 100 kilometres wide, so if the seawater were removed we wouldn't quite be able to see the top of the other side. The side of the trench is perhaps not as steep as we might imagine: the average slope on our way into the trench is about ten degrees. We would certainly notice that incline on foot, but it's not like the scramble we would have had at the rift valley of the Mid-Atlantic Ridge. But the slope on the far side of the trench continues much higher towards the island of Puerto Rico and becomes much steeper, reaching 45 degrees in places. It also has a couple of amphitheatre-shaped underwater embayments in it, about thirty to fifty kilometres wide, which give this slope its name, 'Amphitheatre Escarpment'. These embayments were formed where the shallower sea floor of sedi- mentary rock here eroded away, slipping down the slope in underwater landslides.

Above three kilometres deep the slope flattens out, rising over the last thirty kilometres of our journey to the coast of Puerto Rico. Finally, after meandering for more than 6,000 kilometres across the bottom of the Atlantic, we can poke our head above the waves to see the sixteenth-century walls of the Castillo San Felipe del Morro on the San Juan shoreline ahead. If we were really able to make such an amazing trek, taking in continental shelf and slopes, trudging across abyssal plains and over abyssal hills, climbing seamounts, traversing the Mid-Atlantic Ridge and winding in and out of an ocean trench, I think we would appreciate the true variety of the ocean floor.

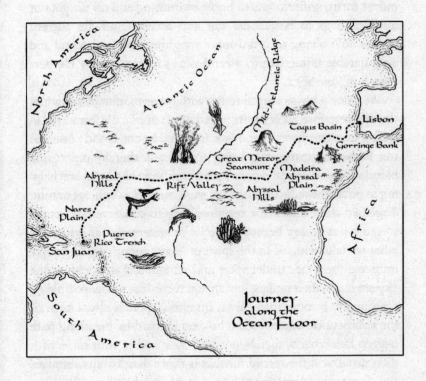

5

How deep can we dive into the ocean without having to protect our bodies from pressure?

In July 2016, New Zealand athlete William Trubridge set a depth record of 102 metres for human breath-hold diving in perhaps its purest form, with no fins to boost swimming and no weights or air-filled bags to hasten descent and ascent. And the current breath-hold diving record using a weighted sled to descend and an inflatable lifting bag to ascend is 253 metres, set by Herbert Nitsch in June 2012.

We share a physiological reflex with other mammals known as the dive response, which is triggered when our face is immersed in cold water. Our heart rate slows, followed by the blood vessels to our limbs and extremities being constricted, thereby prioritising blood supply to critical organs such as the heart and brain and helping to offset the reduced blood pressure from a lower heart output. Together, these automatic reactions help to conserve the oxygen stored in our bodies, because once we're underwater, all we have is what we take with us. In the sport of 'static apnea', where people immerse their face underwater and remain still to conserve the oxygen inside their bodies, the current record for someone holding their breath is more than eleven minutes. There is also a form of the competition where breath-holders prepare by breathing pure oxygen beforehand, thereby loading their bodies with more of it than usual, and the record for that is more than twenty minutes. But despite the surprising potential of our mammalian dive reflex, and the astounding feats of top exponents of freediving, we have to rely on technology to reach the majority of our deep-ocean planet.

We can extend the diving ability of our species by taking an air supply with us, either supplied from the surface through a hose or carried by cylinders of compressed gas in 'rebreather' diving gear and the 'Self-Contained Underwater Breathing Apparatus' (SCUBA) pioneered by Jacques Cousteau and Émile Gagnan. But with any of that equipment, our exposed bodies run into a challenge. Dive just ten metres deep, and the pressure surrounding us becomes roughly the same as that inside a car tyre: double the normal atmospheric pressure that we're used to at the surface. And for every ten metres deeper that we go, the weight of water above us increases the pressure by yet another atmosphere. So at fifty metres deep, the conditions are six times normal atmospheric pressure. To prevent our lungs from getting squeezed by that pressure, personal diving equipment usually supplies us with gas at the same high pressure as the conditions around us. But breathing gases under high pressure introduces four complications (and they are complicated!): oxygen toxicity; the need to decompress safely; inert gas narcosis; and high-pressure nervous syndrome. These problems ultimately limit how far our exposed bodies can go, and prevent us from swimming freely in and out of the deep.

Air contains around 21 per cent oxygen, and although we can breathe pure oxygen at normal atmospheric pressure without immediate problems, too much oxygen can have some toxic effects. The oxygen molecule that we breathe (O_2) is not toxic, but oxygen in our bodies can change into other forms known as 'oxygen radicals', which can disrupt other molecules such as enzymes that our cells need to work normally. Because of those effects, breathing elevated levels of oxygen over many hours can cause problems for our lungs, and for divers, much shorter exposure to oxygen at particularly high pressure can lead to more immediate problems for the brain and central nervous system.

At ninety metres deep, where the pressure is ten times normal atmospheric pressure, the 21 per cent oxygen in normal air becomes equivalent to just over a double dose (200 per cent) of pure oxygen at the surface – a dangerous level that can trigger convulsions because of oxygen toxicity affecting the central nervous system. In the Second World War, Italian frogmen used 'rebreather' diving gear that supplied them with pure oxygen under pressure, which meant they had to stay comparatively shallow or risk running into oxygen toxicity. British frogmen, in contrast, used equipment with a lower percentage of oxygen, allowing them to work at greater depths – and a recommended strategy for dealing with their Italian counterparts was to drag them down deeper, where they would encounter oxygen toxicity but the British divers could remain safe.

If we want to go far beyond the limits of recreational scuba-diving, we need to breathe a mixture of gases that contain a lower percentage of oxygen than normal air for the deep phase of dives. Taking several cylinders containing different gas mixtures to breathe, and carefully switching between them at different depths, makes 'mixed gas' deep diving more technically challenging than recreational scuba-diving, but with care we can potentially overcome the problem of oxygen toxicity.

The air that we breathe also contains around 78 per cent nitrogen, which, unlike oxygen, we don't use – it is an 'inert' gas to our bodies. But although we don't use it, it dissolves in our blood, the fluids that bathe our tissues and the saturated fat of our bodies. High pressure forces more of it into us, and if we then ascend too quickly, the dissolved gas that has built up inside us can come out of solution, like the bubbles that appear when we release the pressure inside a fizzy drink bottle by opening its cap for the first time. Bubbles forming in our body fluids can lead to joint pain, nicknamed 'the bends' after the contorted limb positions that sufferers typically adopt to seek relief. They can also damage our lungs and other vital organs as our bodies try to get rid of the excess gas from our blood. So to avoid the risk of 'decompression sickness', divers ascend slowly, sometimes stopping at particular depths, to give their bodies time to get rid of the gas dissolved under pressure. The current depth record for a single scuba-dive is 332 metres, set by Ahmed Gabr in September 2014 off the coast of Egypt. Gabr spent four years training for his record attempt, and he used different mixtures of gases during his dive. But although his descent to 332 metres took less than fifteen minutes, his return to the surface took more than thirteen hours, because of the stops that he had to make to decompress safely on the way back.

The longer a diver spends breathing gas under high pressure, the longer they usually need to spend decompressing during their ascent afterwards. But if a diver spends a very long time at depth, their body may become fully loaded with dissolved gas – a point known as 'saturation'. At that point, even if they spend more time at depth, the amount of time that they need for decompression afterwards will not increase further. So for repeated dives at great depth, it makes sense for divers to remain under high-pressure conditions between dives, for example staying in a diving bell or pressurised chamber, rather than undergo lengthy decompression periods at the end of each dive. The eventual decompression after a stint of 'saturation diving' can take days or even weeks to complete safely, as divers are gradually reacclimatised to surface conditions. So, with care, we can manage the problem of decompression – but the need for lengthy decompression after deep diving is always there.

If we breathe nitrogen under high pressure, it causes other problems besides the need to get rid of it safely afterwards. 'Inert gas narcosis' describes a range of symptoms from a mild feeling of euphoria and dizziness to disorienting drunkenness, described by Jacques Cousteau's team of pioneering scuba-divers as *l'ivresse des grandes profondeurs*' – 'the rapture of the great depths'. Some of the narcotic problems of high-pressure nitrogen may come from it dissolving in the fatty membranes that coat all of our cells, disrupting the ability of those cells to 'talk' to each other. Transmitting a signal between nerve cells, for example, involves those cells releasing and receiving 'signal' molecules and electrically charged atoms called ions – a perpetual chemical chatter inside us that enables our bodies to perform normally. Some signal molecules attach to large 'receptor' molecules on the outer surface of the membranes that coat our cells, rather like satellite dishes on a roof, and those receptor molecules then trigger a response inside the cell when they receive the signal. Some signal ions also pass into cells through large 'channel' molecules that act like pores in the cell membrane.

The large receptor and channel molecules that our cells use to communicate are like rafts in a sea of much smaller fatty molecules that form our cell membranes. High-pressure conditions can force lots of nitrogen into the fatty 'sea' of the cell membrane, which can in turn play havoc with the receptor and channel molecules floating in it. In some cases, it can squeeze channel molecules shut so that signals cannot pass through them. In other cases, it can switch receptor molecules 'on' or 'off' regardless of whether they are actually receiving a signal or not. But not all the effects of inert gas narcosis result from more of it being dissolved into cell membranes – under high pressure, the gas may affect crucial molecules inside the cells as well.

There are other inert gases that we can breathe instead of nitrogen – that's just the most abundant one that occurs naturally in air. Helium has much less of a narcotic effect under high pressure than nitrogen, so swapping helium for nitrogen as the inert gas – a breathing mix known as 'heliox' – allows divers to go deeper without encountering the same narcotic problems. In 1939, US Navy divers made the first working dives using heliox to salvage the sunken submarine USS *Squalus* from a depth of 73 metres, where nitrogen narcosis had defeated their first dives using ordinary compressed air. But even when breathing heliox, divers exposing their bodies to high-pressure conditions at great depths still encounter another problem, called 'high-pressure nervous syndrome'.

High-pressure nervous syndrome may result from the effects of high pressure on the central nervous system, with symptoms including tremors, uncontrolled eye movements, headaches, fatigue, muscular weakness, nausea, memory problems and problems with coordination. Its onset can be reduced if divers make a very slow descent, taking time to compress in a similar fashion to decompressing on the way back up. But effects can still kick in on very deep dives, regardless of the length of time taken to compress. High-pressure nervous syndrome resembles an overexcitation of the nervous system, whereas inert gas narcosis is more like a stupefaction of normal brain function. So the two problems seem almost opposite in their overall effects.

Dr Peter Bennett, a researcher with a background in anaesthesiology, realised that the narcotic effects of nitrogen might therefore be used to offset the effects of high-pressure nervous syndrome to some extent. Based on that principle, he devised a gas mixture called 'trimix', which combines oxygen, helium and a small amount of nitrogen. In 1981, Bennett oversaw an experiment with a pressure chamber at Duke University in North Carolina that successfully took three divers to a simulated depth of 686 metres using trimix.

Hydrogen gas (H_2), technically known as molecular dihydrogen because it consists of two hydrogen atoms joined together, is the smallest possible molecule in the universe. We can breathe it as an inert gas, and because its molecules are small, it is less dense than nitrogen or helium, which makes it physically easier for us to breathe it in and out of our lungs at high pressure. A mixture of hydrogen and oxygen is not explosive if the oxygen content is less than 5 per cent, so it can be used for the very deep phase of dives, where oxygen content needs to be low to avoid oxygen toxicity. But hydrogen has more of a narcotic effect under pressure than helium, so breathing a mixture of just hydrogen and oxygen cannot take divers deeper than heliox, because of inert gas narcosis. However, a mixture of helium, hydrogen and oxygen – known as hydroheliox – could enable very deep dives as does trimix, using hydrogen's narcotic effect to reduce high-pressure nervous syndrome. Because hydrogen's narcotic effect is weaker than that of nitrogen, the hydroheliox mix needs to contain a greater percentage of hydrogen than the nitrogen percentage used in trimix. But using a high percentage of lighter hydrogen gas makes hydroheliox easier to breathe than trimix under very high pressure, so hydroheliox may be the ultimate gas mixture for extremely deep dives.

The current depth record for divers with unprotected bodies in the open sea is 534 metres, breathing a hydroheliox mixture of 49 per cent hydrogen, 50 per cent helium and 1 per cent oxygen. That record was set by a team of four commercial divers and two French Navy divers in 1988, and the pressure that they experienced was more than fifty times normal atmospheric pressure, or about twenty-five times the pressure inside a car tyre. Working as saturation divers, the team spent eight days compressing inside a chamber aboard their support ship. Thanks to a pressurised diving bell that docked with that cramped accommodation chamber, the divers could then be lowered into the sea for deep dives and return to the chamber aboard ship while remaining at high-pressure conditions. They completed six dives lasting a total of twenty-eight hours between 520 and 534 metres deep, followed by eighteen days decompressing in the chamber back aboard the ship.

In 1992 Theo Mavrostomos made a simulated dive to 701 metres in a pressure chamber ashore, also breathing a hydroheliox mix for the deepest part of the experiment. Mavrostomos's feat remains the record for a human enduring conditions of high pressure on their exposed body. It took Mavrostomos thirteen days to compress to the deep conditions, and twenty-four days to decompress safely afterwards. Even when using hydroheliox to tackle the combined problems of oxygen toxicity, inert gas narcosis and high-pressure nervous syndrome, we cannot avoid the need for lengthy decompression. Safely getting rid of the dissolved gas that builds up in our bodies under pressure is a laws-of-physics limitation, which means we cannot – and probably never will be able to – swim down into the deep ocean and return from it instantly using personal diving gear.

But fortunately, we don't need to. An easier problem, which we can solve through engineering, is how to maintain normal atmospheric pressure inside a sealed undersea vehicle, so that it can take people into the ocean depths without any of these complications. If we remain at normal atmospheric pressure inside such a craft, then there are no problems of oxygen toxicity, inert gas narcosis or high-pressure nervous syndrome, and no need to decompress at the end of a dive. We can dive straight down to much greater depths in a deep-diving submersible, and come straight back, popping open the hatch and immediately climbing out. We might never be able to leave those craft during dives to leave footprints in the soft mud of an abyssal plain, but we can look out of their portholes to witness that landscape firsthand. The challenge is therefore to build a hull that is strong enough to resist the huge difference in pressure between the outside and inside of the vehicle at the bottom of the ocean, and vehicles like that have already taken people all the way to the deepest point in the oceans, more than 200 times deeper than the usual limit of recreational scuba-diving.

6

What is a 'bathynaut', and who were the first bathynauts?

'Bathy' comes from the Greek 'βαθύς' (bathys), meaning deep, and 'naut' comes from the Greek 'ναύτης' (nautes), meaning sailor or voyager. So a bathynaut is someone who voyages deep in the ocean. We usually define the deep ocean as starting at 200 metres, as that's the average depth at the edge of most continental shelves around the world, where the sea floor starts to slope more steeply into the deep. In the open ocean, 200 metres is also taken to be the maximum possible depth where there is enough light for algae to thrive using photosynthesis – the boundary between the surface 'sunlight zone' and 'twilight zone' where daylight dims beneath. So we can define a bathynaut as someone who has dived more than 200 metres deep, beyond the depth of the continental shelf and into the twilight zone. By that definition, bathynauts include people who manage to get that deep using the breath in their lungs, such as record-setting freediver Herbert Nitsch; those who make extremely deep dives using personal diving gear; sailors in naval submarines that cruise that deep; and ocean explorers aboard much deeper-diving Human-Occupied Vehicles.

An alternative definition of a bathynaut might be someone who has dived to one kilometre deep; this is the 'midnight zone' of depths that are forever beyond the reach of sunlight. That would mean the term bathynaut could only be used for people diving in Human-Occupied Vehicles designed for the deep sea, and possibly sailors in some of the most advanced naval submarines taken to their operational depth limit: the K-278 attack submarine of the Soviet Union, for example, had a stronger titanium hull instead of the usual steel and reached a kilometre deep in the Norwegian Sea in 1984, and the USS *Dolphin AGSS-555* submarine was capable of similarly deep dives. But I think a one-kilometre definition would wrongly exclude the first pioneers in Human-Occupied Vehicles designed for the deep sea, whose deepest dive was just short of one kilometre.

Meanwhile, in contrast to the term 'bathynaut', 'aquanaut' has come to describe someone who has spent more than twenty-four hours continuously underwater. That includes divers who stay in an underwater habitat on the seabed, and submariners. There are some bathynauts who are also aquanauts: Dr Sylvia Earle is a great example, having dived deep in mini-subs and led an all-female team of aquanauts who spent two weeks living in an underwater habitat in 1970.

Are there any deep-diving bathynauts who are also astronauts? In 1995 I went to sea on an expedition that included Dr Kathryn Sullivan, who had flown on three Space Shuttle missions and become the first American woman to walk in space. The expedition was led by Dr Cindy Van Dover, herself a pioneer as the first female pilot of the *Alvin* deep-diving submersible, and we were using the US Navy's DSV *Sea Cliff* to investigate hot springs more than two kilometres down on the ocean floor. Dr Sullivan was on the roster to dive in *Sea Cliff*, but poor weather unfortunately prevented her from becoming the first person to go into space and dive deep in a submersible. I heard a rumour in the late 1990s that astronaut Dr Mae Jemison might have made a dive nearly four kilometres deep to the *Titanic* aboard one of the Russian *Mir* submersibles, but I have been unable to confirm that – if it's true, then Dr Jemison would be the first astronaut and deep-diving bathynaut.

There have been far more bathynauts than astronauts, however. It is true that more people have walked on the Moon (twelve so far) than have visited the greatest depth in the ocean (at the time of writing, three people have reached the bottom of the Mariana Trench). But the total number of people who have been into space, defined as 100 kilometres (62 miles) altitude and beyond, is less than 600 at the moment. The number of people who have been into the deep ocean, defined as 200 metres deep and beyond, is in the tens of thousands. And even if we only consider bathynauts who have been deeper than one kilometre, their numbers are still in the thousands, if not more than ten thousand. It is hard to come up with a precise number, because dives in Human-Occupied Vehicles have been so numerous. The *Alvin* submersible is one of the most famous Human-Occupied Vehicles used for science, and since 1964 has made more than 5,000 dives, taking more than 14,000 people into the deep. Japan's *Shinkai 6500* submersible, which carries three people, has completed more than 1,800 dives since 1991. Even allowing for the same pilots and scientists making several dives, those two vehicles have taken far more people into the deep ocean than the number of people who have been into space. And there are, and have been, many more deep-diving submersibles than just those two.

Human-Occupied Vehicles that can take people to 300 metres deep are becoming more common, even being built commercially for the very rich to launch from large private yachts. But there are fewer vehicles that can take people to one kilometre deep, and craft that can carry people more than four kilometres deep are very rare. The *Alvin* submersible, for example, which entered service in 1964, was initially built for dives to just beyond 1.8 kilometres deep. In 1973 a new pressure hull doubled its operational depth limit to around 3.6 kilometres deep; its limit was extended again, to 4 kilometres deep, in 1976, then 4.5 kilometres in 1994 – and upgrading with another new hull is now underway to take it to 6.5 kilometres. But the average depth of all the *Alvin* dives so far is just over two kilometres, in part because of the shallower depth rating of its early years, but also because most of the sites of scientific interest visited by dives, such as undersea hot springs on mid-ocean ridges and long-term experiments established on continental slopes, are at depths shallower than four kilometres.

The number of people who have been more than four kilometres deep in the ocean is therefore much smaller than the number of 'bathynauts' overall, but even then, I think we still outnumber those who have been into space by quite a margin. So at what depth does the number of bathynauts probably match the number of astronauts, at least roughly? I think if we were to consider people who have been more than six kilometres deep, their number might be in the same ballpark as astronauts, in the hundreds rather than thousands. There are, and have been, very few vehicles that can take people to six kilometres deep and beyond. In current service, France's *Nautile* submersible can reach six kilometres; Japan's *Shinkai 6500* is rated to 6.5 kilometres; China's *Jiaolong* has dived to just beyond seven kilometres; and there are also a few military submersibles whose capabilities are not publicly declared. In the past, the Russian *Mir* submersibles, which featured in the film *Titanic*, were rated to six kilometres deep, along with the US Navy's now-retired DSV *Sea Cliff*.

If we consider people who have been to the bottom of the deep ocean trenches, from seven kilometres to nearly eleven kilometres deep, then astronauts finally outnumber the bathynauts. The bathyscaph *Trieste* took two people nearly eleven kilometres deep to the bottom of the Mariana Trench in 1960, and James Cameron's *Deepsea Challenger* returned there in 2012, having also taken Cameron more than eight kilometres deep to the bottom of the New Britain Trench. But the French bathyscaph *Archimède*, which carried three people inside, made the most dives to the bottom of the ocean trenches. In 1962, *Archimède* made eight dives to more than seven kilometres deep in the Kuril–Kamchatka Trench and the Izu–Bonin Trench of the Pacific, including three dives to more than nine kilometres deep. *Archimède* then completed ten dives into the Puerto Rico Trench of the Atlantic in 1964, reaching more than eight kilometres deep, and in 1967 reached 9.75 kilometres deep in the Japan Trench. There is a new Human-Occupied Vehicle touring the deep ocean trenches around the world from late 2018 to 2019, and it may exceed the number of trench dives made by *Archimède*, but even then the number of bathynauts who have been to the bottom of the ocean trenches will still be fewer than the number of astronauts.

The first bathynauts are much less well known than the first astronauts: their names were William Beebe and Otis Barton, and they reached half a mile down in the 1930s aboard their 'bathysphere', which was lowered on a cable rather than being able to manoeuvre on its own like a modern deep-diving submersible. William Beebe was a celebrity naturalist in the 1920s, fêted for best-selling books describing his scientific expeditions for Bronx Zoo. Having investigated marine life around the Galapagos Islands and Haiti, where he used a diving helmet to explore coral reefs, he declared an ambition to study the natural history of deeper waters from an undersea craft of his own devising, as reported by the *New York Times* in 1926. Engineer Otis Barton read about Beebe's ambitions, and already had his own ideas for such a craft, which were rather more advanced than Beebe's – Beebe originally favoured a cylindrical hull, which would not have withstood pressure as effectively as the sphere design that Barton had in mind. Eventually Barton managed to get an audience with Beebe, and Barton's detailed designs and family funding overcame Beebe's initial wariness, leading to a partnership.

In 1929, Barton took his design to the Watson-Stillman Hydraulic Machinery Company in New Jersey, who cast the hull of the bathysphere out of steel. It weighed more than four and a half tonnes, which was unfortunately too heavy to be lifted by the equipment that Beebe had obtained to deploy it from a vessel in Bermuda. So the first hull had to be recast to weigh less than two and a half tonnes, and by the summer of 1930 Beebe and Barton were ready to begin their bathysphere dives off Nonsuch Island near Bermuda.

Barton's final design was a hollow metal ball with an external diameter of almost one and a half metres and a steel wall nearly four centimetres thick. The pioneering bathynauts would crawl inside through a hatchway just thirty-six centimetres across, which would be closed by a steel plate door secured with ten bolts. The door had a raised ridge running around its inner edge that fitted into a shallow groove on the hull, forming an effective seal when packed with some soft white lead. The hull had three portholes that stuck out as tubes from its spherical surface, but only two were fitted with fused-quartz windows, just over twenty centimetres across and more than seven centimetres thick. The other windowpanes built for the bathysphere broke during fitting and testing, so the third porthole-tube was closed off with a steel plug instead.

The whole contraption was lowered on a steel cable from a surface vessel, with the strands of the cable designed to prevent it from spinning. A separate electrical cable provided power for lights and a telephone connection with the world above, passing tightly through a 'stuffing box' from the outside of the hull to the inside. Oxygen was supplied from tanks inside the sphere with automatic valves, and trays of powered calcium chloride and soda lime soaked up moisture and carbon dioxide exhaled by the occupants, who wafted palm-leaf fans to keep the air moving.

After two test dives with the bathysphere empty, Beebe and Barton squeezed through its hatch and on 6 June 1930 ventured to 244 metres deep. Their first dive set a new world record; the previous deepest dive had been to 160 metres deep, in an Alpine lake, using a diving suit. At around 100 metres deep during their first bathysphere dive, water started trickling in at the bottom of the hatch. Beebe gave the order to speed up their descent, reasoning that increasing the pressure on the outside by going deeper would push the 180-kilogram steel hatch more tightly against the hull, preventing any leak from increasing. He was right.

After another test dive with the bathysphere empty and a bit more white lead to fix the leaky hatch, Beebe and Barton attempted another deep descent, but stopped the dive when the bathysphere lost its telephone connection to the surface. But the next day, 11 June 1930, despite an accidental fire aboard the bathysphere's launching vessel overnight they reached 435 metres – the seventh dive of the bathysphere including its empty test dives, the third with people inside and their deepest dive for that year. And five days later, Beebe's colleague Gloria Hollister also set a record as the deepest-diving woman, in a dive to 131 metres. Beebe summed up his first impression of reaching a quarter of a mile deep as follows:

> I shall never experience such a feeling of complete isolation from the surface of the planet Earth as when I first dangled in a hollow pea on a swaying cobweb a quarter of a mile below the deck of a ship rolling in mid-ocean.[5]

The bathynauts returned to Bermuda in the summer of 1932, but the first unoccupied test dive of their bathysphere for that year returned ominously half-full of water. The air squeezed into the top half of the sphere by the water inside was therefore under pressure, so that the centre bolt of the hatch shot across the deck when Beebe cautiously loosened it, 'like a shell from a gun' as he described it.[6] A solid column of water then jetted forth, eventually changing to a geyser-like stream of water mixed with compressed air. 'If I had been in the way I would have been decapitated,' Beebe noted. The leak was traced to the seal of the third porthole, to which the bathynauts had attempted to fit a new quartz window for that year's dives.

On 17 September they attempted another unoccupied test dive, with the third porthole closed off by its old steel plug again. But the tightening of that plug, using only a hand wrench, proved inadequate, and the bathysphere again returned with water and compressed air inside, once more shooting the centre bolt of the hatch across the deck when it was loosened.

After some interruptions from the weather, and two more test dives with the bathysphere successfully emerging watertight, Beebe and Barton on 22 September 1932 set a new depth record of 671 metres. During that dive, the intrepid pair broadcast live on the radio from the bathysphere, describing what they were seeing to listeners across the US via NBC radio and in the UK via the BBC. I like to think of their live broadcast as the first from another world, albeit the hidden face of our own, more than three decades before the televised Moon landings.

The final series of bathysphere dives took place in 1934, with funding from the National Geographic Society, whom Beebe had promised they would reach half a mile down. The bathysphere had been overhauled for the challenge, for example with new quartz windows, a better-designed stack of trays of chemicals to soak up moisture and carbon dioxide and an electric blower to replace the palm-leaf fans. In fact, Beebe remarked, 'of the old bathysphere which had carried us down and up so safely nothing remained save the steel skeleton itself.'[7] But a test dip in St George's Harbor, with Beebe and Barton inside, was cut very short when water poured in around the hatch, which for such a shallow dive had only been secured with four bolts instead of ten. Having rectified that oversight, they continued their rehearsal dive of the renewed bathysphere, lowering it about seven metres down to the soft mud of the harbour bottom while the deck crew practised their procedures.

The next day the bathysphere made a flawless unoccupied deep dive, and on 11 August Beebe and Barton crawled aboard for their first proper dive of the year, beating their former record by reaching 765 metres. Then on 15 August, they passed their half-mile-down target and reached 923 metres deep – a record that would stand for another fifteen years. Peering through the quartz glass portholes, Beebe and Barton became the first people to watch deep-sea animals in the ocean, alive with their squirts and flashes of bioluminescence, rather than studying dead or dying specimens hauled up in nets. Beebe wrote a popular book, *Half Mile Down*, which contains a vivid account of witnessing our planet's 'inner space':

> Whenever I sink below the last rays of light, similes pour in upon me . . . The eternal one, the one most worthy and which will not pass from mind, the only other place comparable to these marvelous nether regions, must surely be naked space itself, out far beyond atmosphere, between the stars, where sunlight has no grip upon the dust and rubbish of planetary air, where the blackness of space, the shining planets, comets, suns, and stars must really be closely akin to the world of life as it appeared to the eyes of an awed human being, in the open ocean, one half mile down.[8]

After the Second World War, Barton continued to pursue his deep-diving ambitions, and he designed a new craft called a benthoscope with engineer and fellow Harvard alumnus Dr Maurice Nelles. Like the bathysphere, the benthoscope still dangled from the surface by a cable, but rather than swinging around at mid-water depths, it was designed to observe the seabed, trundling over it on wheels almost as wide as the sphere itself. On 19 August 1949 Barton set a new depth record by descending to 1,372 metres aboard the benthoscope off the coast of California.

Beebe and Barton were the first to venture into the deep, and both are among my heroes of ocean exploration. Beebe was a great writer as well as a naturalist, and he once wrote that 'There are two kinds of thrill in science. One is the result of long, patient, intellectual study ... But the other thrill lies in a completely unexpected discovery'.[9] We're still making completely unexpected discoveries, and feeling that same thrill, alongside our patient intellectual study in the deep ocean today. And Barton summarised the experience of exploring the deep ocean so beautifully in his book *The World Beneath The Sea*, published in 1953: 'No human eye had glimpsed this part of the planet before us, this pitch-black country lighted only by the pale gleam of an occasional spiralling shrimp.'[10]

At the start, bathynauts were not a boys-only club. After becoming the deepest-diving woman in 1930, Gloria Hollister went deeper as Beebe's colleague aboard the bathysphere in 1934, setting a new record of 368 metres and thereby going beyond the 200-metre definition of a bathynaut. Hollister subsequently led scientific adventures above the waves, for example exploring 200 miles of jungle in Central America in 1936, including flying in a small plane to discover forty-three otherwise inaccessible waterfalls. But there was a hiatus in the mid-twentieth century for female bathynauts, maybe caused by ocean exploration crossing paths with the Cold War and the military traditions that excluded women from diving in submarines. As far as I'm aware, the first female bathynaut to dive more than one kilometre deep was Dr Ruth Turner, who on 13 August 1971 became the first woman to dive in the *Alvin* submersible to pursue her research in deep-sea biology. That dive reached a depth of 1,829 metres, and it was the first of forty-seven dives that she made in *Alvin* over the next nineteen years.

Since the 1970s, there have been many notable female bathynauts. Dr Sylvia Earle made a solo dive in an armoured diving suit at 381 metres deep off the coast of Hawaii in 1979, and subsequently made a solo dive to 1,000 metres in the *Deep Rover* submersible off the coast of California in 1985. Dr Edith Widder also dived

alone in an armoured diving suit in 1984, to observe the bioluminescence of deep-sea life at more than 260 metres deep, and has
since made many more deep dives in subs to study that phenomenon further. In 1990, Dr Cindy Van Dover qualified as the first
female pilot of the *Alvin* submersible, and subsequently led forty-
eight dives as pilot-in-command, in addition to her many more
dives in submersibles as a scientist. Today there are female
submersible pilots from Japan and China routinely leading dives
into the ocean depths, along with female scientists investigating
the deep ocean aboard submersibles. Women are also now the
'deepest divers' of several countries: for example, Trinidad's
deepest-diving bathynaut is Dr Diva Amon, one of my former
research students.

7

Was there an 'inner space race' to send people to the greatest depths of the ocean?

In a sense there was an 'inner space race', but it was not quite like the rivalry between Cold War superpowers that gave us the 'space race' of cosmonauts and astronauts into orbit and beyond, and its story is far less well known. When it came to sending people to ever-greater depths in the oceans, the contender nations were Belgium, France and the United States – and weaving between them was a brilliant Swiss physicist-adventurer, and his equally brilliant son.

The early bathysphere and benthoscope were like upside-down tethered balloons, capable only of going up and down by winding in or paying out their cable, similar to the early tethered balloons of the Montgolfier brothers. The next step in the evolution of deep-diving vehicles was a transition from 'tethered balloon' to something more like an underwater 'airship', no longer attached to a surface vessel and capable of manoeuvring itself horizontally. And it took a balloonist to make that transition: in the late 1930s, veteran balloon designer Auguste Piccard devised a new undersea craft that he called a 'bathyscaph'.

Swiss-born Piccard was a professor of physics in Belgium, and in the early 1930s he designed a pressurised gondola for high-altitude balloon flights, in which he set records culminating in 1932 with an ascent to an altitude of 16,936 metres to measure cosmic rays. The gondola from Piccard's ascent was exhibited at the Century of Progress Exposition in Chicago in 1933, with Beebe and Barton's bathysphere on show beneath it. Piccard realised that the design principles of his pressurised capsule could be adapted to make a craft that could resist the pressure of the abyss, and the principles of an airship could be adopted to make that undersea craft free-swimming, instead of dangling from a surface vessel by a cable.

In place of the bag of hot air or lighter-than-air gas that a balloon or airship uses to float in air, Piccard's 'bathyscaph' had a large tank of petrol – a liquid lighter than water. Below this buoyancy tank, just like the gondola of a balloon, was the capsule that carried the human crew – another hollow metal ball, similar to the bathysphere. Rather than being lowered down and hauled up on a cable, the untethered bathyscaph was loaded with iron shot as ballast weight at the surface, so that the craft sank like a stone. On nearing the seabed, the occupants could jettison the iron shot ballast to slow their descent, bringing the craft to rest just above the sea floor. The bathyscaph could then scoot around at the sea floor using a small propeller and rudder. To ascend after working at the ocean floor, the crew could release further ballast weight, allowing the craft to float back up to the surface thanks to the buoyancy of its large tank of petrol.

Piccard's record-setting balloon had been funded by the Belgian government's Fonds National de la Recherche Scientifique, and named *FNRS-1* after that organisation. He successfully approached the same organisation in 1937 for funding to build his first bathyscaph, which he eventually named *FNRS-2*. In the late 1930s he built models and tested them to pressures that simulated depths greater than the deepest ocean trenches, and he found that a new acrylic safety glass would be ideal for porthole windows, as quartz windows previously used in the bathysphere would not be strong enough at much greater depths. The Second World War interrupted his progress, but by 1948 the *FNRS-2* was ready to go to sea – and by that time Piccard's son Jacques, now in his mid-twenties, had also become his colleague in bathyscaph development and operation.

For the first dives of the *FNRS-2*, Piccard chose deep waters close to the Cape Verde Islands off the west coast of Africa, and the Belgian government provided an aging freighter called *Scaldis* to carry the bathyscaph. The expedition was planned as a collaboration with French scientists and the French Navy – in particular, its Undersea Research Group founded by Jacques Cousteau. With a base in nearby Dakar, the French Navy provided reconnaissance and rescue aircraft, two frigates and the *Élie Monnier* of the Undersea Research Group. The plan was to make a series of dives to around four kilometres deep, with Cousteau included in the list of those who would be diving with Piccard.

A shallow test dive on 26 November 1948, with Auguste Piccard and French naturalist Théodore Monod inside, was successful. A subsequent unoccupied deep dive, with a timer set to release the final ballast weight automatically, achieved its target depth of 1,400 metres successfully. But on its return the craft ran into problems at the surface, as the thin plating of the buoyancy tank proved inadequate to withstand the ocean waves. The crane of the *Scaldis* could only lift the bathysphere back aboard when empty of petrol, and in the rough seas the crew could not attach hoses to siphon off the petrol in the buoyancy tank. The only course of action to allow the craft to be lifted out of the water and prevent further damage from the waves was to dump all of the petrol out of the tanks into the sea. After a tense night completing that task, the bathyscaph was hoisted aboard, but in need of repair and with no more petrol available for further dives. The expedition had to be cut short without taking people into the deep. Cousteau, a first-hand witness, describes the build-up and final drama of the *FNRS-2* in a chapter of his book *The Silent World* – another essential volume for any devotee of ocean exploration, alongside Beebe's *Half Mile Down*.

The Belgian government were initially unwilling to fund further work, but after negotiations with the French Navy, a deal was reached in 1950 involving some Belgian funding, the use of the French naval base at Toulon and Auguste Piccard as a consultant in rebuilding the bathyscaph, which would now be called *FNRS-3*. But Piccard found that arrangement frustrating, as the French Navy seemed set on making their own progress without him, so he left them to it.

Meanwhile, son Jacques had been raising funds in Switzerland and courting industrialists in the city of Trieste in Italy. By 1952 the Piccards had secured enough funding to build a second bathyscaph in Trieste, and the new craft would bear that city's name as its own. In September 1953 the bathyscaph *Trieste* was complete, and the father-and-son crew of Auguste and Jacques climbed aboard to dive 3,167 metres deep near Naples in the Mediterranean – a new record, more than twice the depth reached by Barton in his benthoscope four years earlier. And with that dive, at the age of sixty-nine, Piccard senior became the first person to set records for going 'up' in his balloon and 'down' in his bathyscaph.

The French Navy were hot on their heels with the *FNRS-3*, however, and on 15 February 1954, Georges Houot and Pierre Willm set a new depth record off Dakar, reaching 4,050 metres in the rebuilt craft that Piccard had originally designed. The *FNRS-3* continued in service with the French Navy until the 1960s, and now adorns the waterfront in the park of the Tour Royale fort in Toulon on the southern coast of France.

Despite having proved the capability of the *Trieste*, the Piccards ran into another shortage of funds for further work. But by the late 1950s, some scientists were becoming aware of, and enthusiastic about, the potential of deep-diving vehicles for investigating the depths of the ocean. Marine geologist Dr Robert Dietz was one of them, having started using scuba equipment to make *in situ* observations of underwater geological features, and dived to 120 metres deep in the Japanese diving bell *Kuroshio* in 1952. While serving with the US Office of Naval Research in London, Dietz met Jacques Piccard, and invited him to give a presentation about the bathyscaph *Trieste* at a meeting of 103 scientists in

Washington DC on 29 February 1956. That meeting led to a charter of the *Trieste* by the Office of Naval Research for a series of twenty-eight dives in the Mediterranean in 1957 that reached 3,200 metres deep, with Jacques Piccard as pilot and US scientists as passengers. Impressed by its performance, the US Navy then bought the *Trieste* outright for US$200,000 in 1958, retaining Jacques as its pilot.

The US Navy privately set an ambitious target, known as Project Nekton, for their newly acquired bathyscaph: to reach the deepest point in the oceans, the Challenger Deep of the Mariana Trench, nearly eleven kilometres beneath the waves. In 1959 the *Trieste* was refitted for that goal, with a stronger and slightly larger personnel sphere cast by Krupp in Germany, a larger buoyancy tank and greater ballast capacity. And on 15 November 1959 Jacques Piccard and Dr Andy Rechnitzer, chief scientist of Project Nekton, set a new depth record with a dive to 5,532 metres while testing the refitted vehicle near Guam in Micronesia, which would be the base for attempting the Mariana Trench dive.

The new sphere of the *Trieste* encountered a problem during that deep test dive, however. It was cast in three pieces, which were glued together by epoxy resin. The metal of the sphere became cold during six hours in the deep, and on returning to warm tropical surface waters, thermal expansion across the glue joints slid the pieces of the hull fractionally out of alignment with each other and allowed some seawater to seep into the sphere. Back in Guam, therefore, mechanical straps were devised to hold the pieces of the hull together, with rubber strips underneath them bonded over the joints of the hull with the same sealant used for gaskets in car engines. The pressure of the ocean, the team reasoned, would then help to keep everything in place during a dive. On 15 January 1960 Jacques Piccard and Lieutenant Don Walsh, officer-in-charge for Project Nekton, made a preparatory dive to 7,010 metres in the Nero Deep of the Mariana Trench.

Then on 23 January 1960 the *Trieste* set the ultimate record, descending 10,916 metres to the Challenger Deep, again with Piccard and Walsh aboard.

To avoid the problems that the *FNRS-2* had encountered through being carried by a ship and craned in and out of the ocean, the *Trieste* was towed to and from its dive sites, with the crew then boarding and disembarking via a small boat. After towing the *Trieste* more than 300 kilometres out from Guam to the Challenger Deep, Piccard and Walsh boarded despite swelling seas, climbed down the access tunnel through the buoyancy tank and into the pressure sphere below and sealed its hatch at around 8 a.m. Shortly afterwards, just before 8:30, the *Trieste* started to dive. The need to complete the dive during daylight hours, to avoid having to secure the bathyscaph in darkness and High Seas for towing back to Guam, provided a ticking clock.

Near the surface, the *Trieste* encountered quite strong thermocline layers, where the temperature of the water drops and therefore its density increases, slowing the descent of the bathyscaph. By releasing some of the petrol in the buoyancy tank, Piccard and Walsh were able to sink through these layers and try to speed up their descent to keep on schedule. At around 9,450 metres deep, the vessel was jarred by a muffled bang. Nothing untoward was apparent inside the sphere, so the dive continued, but later inspection by Walsh found that a curved perspex window had cracked on the external access tunnel outside the hatch of the sphere. That access tunnel filled with water during dives, so there was no immediate threat to the occupants from the crack. But if the window broke completely while pumping out the access tunnel with compressed air as usual after the dive, it would be very difficult for Piccard and Walsh to disembark from the bathyscaph. Fortunately, it held.

At 1:06 p.m., the *Trieste* coasted to a halt on the soft mud bottom at the deepest place in the oceans, after descending for just over four and three-quarter hours. Piccard and Walsh were able to spend twenty minutes there before they needed to leave to get back to the surface on time, but they could not take any photographs at the sea floor because the fine silt stirred up by the craft took longer than twenty minutes to clear. They did, however, see some marine life scurrying about through the haze, including what they described as a foot-long, whitish 'flatfish' like a sole. Marine biologists still debate that observation today, as fish are not thought to reach that deep, although other animals such as sea cucumbers can. Around 1:24 p.m. Piccard and Walsh released ballast to leave the bottom, returning to the surface over three and a half hours. With the access tunnel slowly blown dry to avoid further damage to its cracked window, the pair eventually emerged to rougher seas than those they had left. The *Trieste* was hitched up for towing and returned to Guam in triumph, with the US Navy announcing its success to the world.

Jacques Piccard returned to Switzerland after Project Nekton, and designed new underwater vehicles. His father passed away in 1962, and Jacques named one of his new creations *Auguste Piccard* after him: a tourist submarine built for the 1964 Swiss national exhibition, which could carry forty passengers at a time and took around 33,000 people to more than 100 metres deep in Lake Geneva during the year-long exhibition. Jacques Piccard also designed the *Ben Franklin* 'mesoscaphe' for long-duration missions drifting in the ocean with six people aboard. In 1969, at the same time that the *Apollo 11* astronauts made their journey to land on the Moon, the *Ben Franklin* undertook a month-long undersea journey with Jacques Piccard and five other bathynauts aboard, covering more than 2,200 kilometres from Florida to Nova Scotia at an average depth of nearly 200 metres and dipping down as far as 550 metres, to study the Gulf Stream and the seabed.

Walsh prepared the *Trieste* for further service with the US Navy, and worked on plans for a *Trieste II* that entered service in 1964, but both were given an operational depth rating of just over six kilometres by the US Navy and never dived to the bottom of an ocean trench again. The French Navy led subsequent human exploration of the deep ocean trenches with their bathyscaph *Archimède*, which was designed by Pierre Willm and launched in 1961. *Archimède* completed several campaigns of ocean trench dives, and also contributed to scientific exploration outside trenches, including taking people to the deep mid-ocean ridge for the first time, before being decommissioned in the 1970s. The bathyscaph *Archimède* is currently an exhibit at the Cité de la Mer in Cherbourg on the northern coast of France.

Between 1960 and 2012, four vehicles including the *Trieste* visited the Challenger Deep of the Mariana Trench: two Human-Occupied Vehicles, and two Remotely Operated Vehicles. Thirty-five years after Piccard and Walsh's dive, the next visitor to Challenger Deep was Japan's Remotely Operated Vehicle (ROV) *Kaikō*, which appropriately means 'trench'. *Kaikō* was not originally conceived with trench exploration in mind, but as a potential rescue vehicle for Japan's *Shinkai 6500* Human-Occupied Vehicle. Although the *Shinkai 6500* has an operational depth limit of 6.5 kilometres, its hull is designed to withstand maximum pressure conditions of just over ten kilometres, so the reach of its rescue vehicle needed to match that, in case the sub should ever sink that deep with people inside. But the designers soon realised that the same capability could be used to investigate ocean trenches.

On 24 March 1995, *Kaikō* reached 10,911 metres deep in the Mariana Trench. The ROV was then used in science expeditions investigating the bottom of the trench, completing nineteen dives there by 1998. As well as filming the bottom of the trench, *Kaikō* collected samples including sea-floor sediments and a 4.5-cm-long amphipod crustacean. The dives didn't encounter any fish, despite Piccard and Walsh's report of seeing one in 1960. *Kaikō* was used for many other science dives elsewhere, but was unfortunately lost in the South China Sea on its 296th dive in

2003, as Typhoon Chan-hom closed in towards the research ship RV *Kairei*.

On 31 May 2009, the *Nereus* vehicle operated by Woods Hole Oceanographic Institution in the US became the third vehicle to reach the bottom of the Mariana Trench. *Nereus* was a new kind of 'hybrid' vehicle, combining Remotely Operated Vehicle (ROV) and Autonomous Underwater Vehicle (AUV) technology into an efficient and cost-effective tool for reaching beyond six kilometres deep. Like a conventional ROV, *Nereus* could be directly controlled from the surface, via a thin fibre-optic tether. But like an AUV, *Nereus* carried its own power in batteries, removing the need for a heavier conducting cable that usually powers an ROV from the ship above. After proving its capability at the Challenger Deep, *Nereus* went on to dive elsewhere, for example exploring hydrothermal vents in the Cayman Trough of the Caribbean. But *Nereus* was unfortunately lost on 10 May 2014 while exploring the Kermadec Trench at a depth of 9.99 kilometres, most likely after an implosion of one of the air-filled glass spheres that it carried for buoyancy.

The next human visitor to the bottom of the Mariana Trench was Hollywood director and ocean explorer James Cameron, on 26 March 2012. Cameron was the sole occupant of his *Deepsea Challenger* vehicle, which reached the bottom of the Mariana Trench in two hours thirty-seven minutes – nearly twice as fast as the *Trieste* – thanks to a 'vertical torpedo' design. Overall, the *Deepsea Challenger* weighed less than a tenth in air than the *Trieste*, with pressure-resisting glass foam providing buoyancy instead of the large petrol tank of a bathyscaph. The personnel sphere into which Cameron folded his 1.87-metre tall body had an internal diameter of 1.09 metres, and a steel wall 6.4 centimetres thick. Cameron spent three hours at the sea floor of the Challenger Deep, before ascending to the surface in just seventy minutes. The *Deepsea Challenger* vehicle did not dive again, however, and while its innovative cameras and lights were used on other vehicles by Woods Hole Oceanographic Institution, its hull was irreparably damaged by a fire in July 2015 while being transported on the back of a truck.

A company called Triton Submarines has recently built and tested a new titanium-hulled submersible – the *Triton 36000/2* – that can carry two people to the bottom of ocean trenches, and is setting out on a voyage that will visit the deepest points in all five oceans, including dives at Challenger Deep in 2019. Japan and China have also declared ambitions to build Human-Occupied Vehicles capable of reaching the bottom of the deep trenches, so human visits to the furthest reaches of our oceans may become more routine in future.

'The March of the Bathynauts'
Deepest Dives

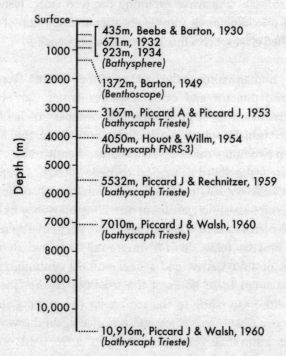

Depth (m)	
Surface	435m, Beebe & Barton, 1930
	671m, 1932
1000	923m, 1934
	(Bathysphere)
2000	1372m, Barton, 1949
	(Benthoscope)
3000	3167m, Piccard A & Piccard J, 1953
	(bathyscaph Trieste)
4000	4050m, Houot & Willm, 1954
	(bathyscaph FNRS-3)
5000	
	5532m, Piccard J & Rechnitzer, 1959
6000	(bathyscaph Trieste)
7000	7010m, Piccard J & Walsh, 1960
	(bathyscaph Trieste)
8000	
9000	
10,000	
	10,916m, Piccard J & Walsh, 1960
	(bathyscaph Trieste)

8

How did modern deep-diving vehicles evolve from early bathyscaphs, and how do they work?

Outside the deep ocean trenches, Human-Occupied Vehicles evolved rapidly, spurred by the enthusiasm of some scientists for exploring the deep first-hand and by the exigencies of the Cold War. Soon after Project Nekton, the refitted *Trieste* and successor bathyscaph *Trieste II* were needed to investigate one of the worst submarine accidents in naval history. On 10 April 1963, the US Navy's nuclear submarine USS *Thresher* was lost with 129 people aboard during test dives after a refit, about 350 kilometres offshore of Cape Cod in the Atlantic. The *Trieste* found and photographed wreckage at 2,600 metres deep, and the *Trieste II* subsequently recovered some of it. The *Trieste II* also photographed wreckage of the USS *Scorpion*, another nuclear submarine lost in the Atlantic in 1968, at a depth of more than 3,000 metres about 740 kilometres south-west of the Azores.

The French bathyscaph *Archimède* similarly took part in investigating submarine accidents, searching unsuccessfully for the wreck of France's *Minerve* diesel-electric submarine, which was lost in the Mediterranean in 1968, and investigating the wreckage of the *Eurydice*, lost in 1970. But the bulky buoyancy tanks of the bathyscaphs made them quite slow and cumbersome at the sea floor, with limited ability to manoeuvre around wreckage to pick up items. As Robert Dietz noted after diving in the *Trieste* in 1957, 'A basic restriction on the maneuverability of the Trieste was imposed by the large size of the gasolene-filled float. It would seem to me that future deep-sea craft should try to dispense with such cumbersome floats.'[11]

In 1959, Jacques Cousteau's team launched their 'diving saucer', which could carry two people and had an operational depth of 400 metres. With a lighter aluminium hull instead of steel, it did away with a bulky buoyancy tank of petrol, instead using water-filled tanks that could be pumped out and replaced with compressed air for buoyancy. But using that approach for vehicles to go much deeper would be problematic, requiring an air supply stored under extremely high pressure to force water out of such tanks at greater depths. Fortunately, in 1955 the Bakelite company created a new kind of material that it named 'syntactic foam'. This material consisted of tiny air-filled glass spheres, set in a hard resin such as epoxy. With careful tweaking of the ingredients, it could potentially resist high pressure without being compressed and losing buoyancy from the 'foam' of tiny air-filled spheres inside it. And because its buoyancy comes from the air that it contains, syntactic foam is far less bulky than a tank of petrol with the same effect, and it has been used in deep-sea submersible construction ever since.

Allyn Vine from Woods Hole Oceanographic Institution was also at the 29 February 1956 meeting in Washington DC where Jacques Piccard had presented details of the *Trieste* and US scientists discussed future technology to explore the deep oceans. Vine and a few colleagues at Woods Hole were already convinced of the potential value of vehicles that could carry people to study the deep ocean first-hand, and by 1962 they had secured funds to build such a craft, capable of carrying three people to a depth just beyond 1,800 metres. The new vehicle was designed by Harold Froehlich of the General Mills Corporation, which won the contract to construct it. The submersible was named *Alvin*, a portmanteau of 'Al' and 'Vin' to acknowledge Allyn Vine's role in its creation, and was commissioned into service at Woods Hole Oceanographic Institution on 5 June 1964.

Alvin was smaller and more manoeuvrable than a bathyscaph thanks to the use of syntactic foam for buoyancy, and provided a blueprint for the fundamental features of most modern deep-diving submersibles. Although developed for scientific investigation of the deep ocean, the new submersible immediately saw Cold War service with the US Navy as well. As a task for its test dives during 1965, *Alvin* inspected the navy's Project Artemis underwater listening array in deep water off the coast of the Bahamas. Then on 17 January 1966, a mid-air refuelling accident resulted in the crash of a B-52G bomber over the coast of Spain, and one of its four hydrogen bombs splashed down into the Mediterranean Sea. *Alvin* found the bomb on the seabed, but the first attempt to recover the weapon failed, returning it to the deep. Two weeks later, *Alvin* found the bomb again, and it was eventually brought to the surface by an early Remotely Operated Vehicle (ROV), which became entangled in the bomb's parachute but was nevertheless able to haul it up.

Also commissioned in 1964, and working alongside *Alvin* in the search for the hydrogen bomb, was another new deep-sea vehicle called *Aluminaut*. With a cylindrical hull nearly sixteen metres long, *Aluminaut* was much more like a conventional submarine in shape, and could take six people inside to an operational depth limit just beyond 4,500 metres. Its hull was made from aluminium, which allowed it to be buoyant without the need for additional materials, and *Aluminaut* had particularly sophisticated manipulator arms for working at the sea floor. Those proved useful after an accident involving *Alvin* on 16 October 1968, when the cables hauling it out of the ocean onto its mothership broke, and the personnel sphere was swamped by the sea when it hit the water. The crew scrambled out safely, but *Alvin* sank to the sea floor around 1,500 metres below. Almost a year later, in September 1969, *Aluminaut* attached lines that hauled the water-filled submersible back to the surface. But *Alvin* ultimately outlasted the larger *Aluminaut*, which went out of service in 1970.

When *Alvin* was being built, two steel pressure spheres were created in addition to the one used in the final vehicle. Those other two spheres became the hulls of two sister-subs of similar design to *Alvin*: DSV *Turtle* and DSV *Sea Cliff*, both operated by the US Navy. DSV *Sea Cliff* eventually received an even stronger titanium sphere that allowed dives to 6,000 metres deep, surpassing *Alvin*. *Turtle* and *Sea Cliff* are now retired, but they were occasionally used for civilian science as well as naval duties. *Alvin* is still going strong as a workhorse for deep-sea science, and is currently undergoing another upgrade that will increase its capability even further.

Since the 1970s, the line-up of other notable Human-Occupied Vehicles taking people more than one kilometre deep has included: France's *Cyana* (now retired) and *Nautile*; the *Pisces III* to *XI* series of subs built in Canada and used by various organisations and companies around the world; Russia's *Mir-1* and *Mir-2* submersibles; Japan's *Shinkai 2000* (now retired) and *Shinkai 6500*; and China's *Jiaolong*, which entered service in 2010.

In addition, various military deep-submergence vehicles still lurk in the shadows. For example, a previously classified Russian sub nicknamed *Losharik* was revealed to the world in 2012. The *Losharik* is reportedly capable of sustaining a crew of twenty-five for several weeks in operations deeper than 2.5 kilometres and perhaps down to six kilometres, with a hull of multiple titanium spheres and a nuclear reactor. *Losharik* came to light when it was used to collect data and geological samples in the deep Arctic Ocean for the claim submitted by Russia to the United Nations to increase its sea-floor territory towards the North Pole, where there are potentially rich reserves of oil and gas. The US Navy also previously had another small nuclear-powered vehicle, called *NR-1*, for long-duration deep-sea missions down to more than 700 metres deep, which was equipped to creep across the sea floor on wheels and was occasionally used for civilian science.

The eponymous shape of Beebe and Barton's bathysphere lives on in many deep-diving submersibles, because a sphere is the most effective shape to distribute pressure evenly around a hull. So although many modern submersibles look 'submarine-shaped', typically there is a hollow ball somewhere at their heart, cocooning the occupants from the pressure of the deep. The rest of the sub consists of components such as batteries, hydraulics and buoyancy materials, all wrapped up in an outer cladding that is not actually a pressure hull.

For modern submersibles that dive more than a kilometre deep, the pressure sphere is often made of a steel alloy or titanium – the latter is particularly strong yet light, but can be difficult to work with. The personnel sphere is a piece of precision engineering: the titanium hull of Japan's *Shinkai 6500* submersible, for example, is 73.5 millimetres thick, enclosing a space two metres across inside it. If it were not perfectly spherical by half a millimetre across its overall 2,147-millimetre external diameter, the pressure would not be distributed evenly around the outside of the hull, with potentially catastrophic consequences for the bathynauts inside. And wherever something inside the pressure hull needs to be connected to something outside, such as the controls for external manipulator arms, lights and cameras, there are 'penetrators' that carry the electrical connections through the hull, and these need to be engineered very carefully to withstand pressure as effectively as the hull itself.

With a metal hull, the only direct view of the world outside is through a few small portholes, fitted with acrylic windows shaped like truncated cones so that pressure pushes them in hard against the hull. But for shallower submersibles with an operating depth of a kilometre or less – and a few that can go deeper – the whole personnel sphere can be made out of transparent acrylic instead. There are a few other panels in that hull, for example to mount the hatch of the vehicle and the penetrators that carry connections between the inside and outside of the hull, but an acrylic sphere provides a superb panoramic view. The *Deep Rover 2* submersible – one of many marvellous undersea vehicles designed by engineer Graham Hawkes – has an acrylic sphere that accommodates two people in comfort with proper seats, as does the front compartment of the two *Johnson Sea Link* submersibles, and I have enjoyed many hours aboard them feeling like I was inside a bubble suspended in the ocean.

A few submersibles have different-shaped pressure hulls made from other materials. The 'diving saucer' of Jacques Cousteau's team had a flattened elliptical hull in which its occupants lay horizontally, made from two sections of aluminium and capable of operating to 400 metres deep, and aluminium also allowed the *Aluminaut* of the 1960s to have a more normal submarine-shaped hull and dive much deeper. More recently, the OceanGate company has built a five-person submersible with a tubular-shaped hull capable of operating at half a kilometre deep, and also created a similar hull with carbon-fibre to operate at four kilometres deep.

Life support inside a modern deep-diving submersible has changed little in principle since Beebe and Barton's bathysphere. The oxygen that we consume is replaced from cylinders, and the carbon dioxide we produce is mopped up by chemical 'scrubbers', ventilated by fans. The remainder of the air – the 78 per cent nitrogen that our bodies don't use, for example – is just recirculated.

Syntactic foam provides a great buoyancy material for modern deep-diving submersibles, and can even be used as a structural material in advanced designs, such as James Cameron's *Deepsea Challenger*. Some deep-diving submersibles still use ballast weights to help them dive down into the ocean – *Alvin*, for example, has metal drop-weights that it jettisons and leaves behind at the ocean floor – but others, such as the *Mir* submersibles, use seawater ballast tanks. At the sea floor, pilots can adjust the tilt of a vehicle, for example to compensate for heavy rock samples collected in the basket at the front of the sub, by using a system of trim tanks. Sometimes these contain liquid mercury that can be moved around between them, but seawater ballast tanks can sometimes double up as a trim-tank system as well.

To manoeuvre, deep-diving submersibles have 'thrusters' consisting of a motor driving a propeller inside a tunnel-like shroud, with many now favouring several individual thrusters mounted in different directions rather than a larger main propeller that swings around on a rudder-like moveable mount. Piloting may involve a joystick with a control system that translates its movements into commands for different thrusters, but some subs just have a box with switches for each individual thruster, with pilots learning combinations of switches that become instinctive to move the sub in any desired direction. In addition to being able to move forwards and backwards, and turn, subs with multiple thrusters can move sideways like a crab without turning, and with the right combination of thrusters even circle around a target while keeping their bow pointing towards it.

At the front of a modern sub, there are usually manipulator arms for picking up objects and deploying equipment, and these can be either electric or hydraulic, with the latter typically giving greater lifting strength but perhaps less precision. Each arm has several moveable joints, some bending like our elbows and some rotating like our wrists, with a claw-like grabber at the end of the arm that can open and close. Each moveable joint in the arm is called a 'function', so a 'seven-function' manipulator arm has more moveable joints and may therefore be more dexterous than a 'five-function' arm. Controlling one of these arms is a skill that takes practice: the control unit typically looks like a miniature model of the arm, but the control of each joint is switched on and off individually, so how the whole controller looks when you move part of it is not necessarily what the actual arm is doing.

Also at the front of the vehicle is the 'sample basket' that carries any equipment for the arms to deploy, and has containers for anything that the arms collect, such as rock samples. Planning the configuration of the sample basket is an important part of preparation for each dive, making sure the necessary equipment for that dive's tasks is accessible and that there is space and possibly specialist containers for anticipated samples, within overall payload limitations. Equipment that a sample basket might carry includes a 'slurp gun': a suction sampler, like an underwater vacuum cleaner, that can suck up specimens of animals for biologists, collecting them into individual water-filled chambers that rotate on a carousel to keep different samples separate. One of the manipulator arms picks up the nozzle of the slurp gun to wave it over whatever a scientist wants to collect, while an operator switches the suction pump on and off, and then the arm drops the nozzle back into a holster on the sample basket, ready to use again. Most equipment, such as the slurp gun nozzle, is fitted with T-shaped metal handles that the claws of the manipulator arms can pick up.

In recent years most submersibles have used HMI (hydrargyrum medium-arc iodide) external lights, which are typically used in the film and TV industry, as they are more power-efficient than other types and produce light with some similar characteristics to daylight. But some vehicles have now started to use LED (light-emitting diode) lights, as they are even more power-efficient, though the quality of their light is perhaps not quite the same. Most Human-Occupied Vehicles carry an external video camera on a pan-and-tilt mount controlled and monitored from inside the sub, and often a photo camera as well. For filming documentaries, submersibles can be rigged with extra lights on long boom arms to get the best lighting for cinematic images. Taking that idea even further, some of the deep-sea submersible footage shot specifically for the BBC's *Blue Planet II* series involved two submersibles diving together, with the second vehicle helping to light some shots from additional angles.

All of the sub's thrusters, manipulator arms and lights need power, which comes from on-board batteries. Battery technology has improved rapidly in recent years, driven by the demand for ever-slimmer units with longer-lasting performance in personal devices such as smartphones. Some Human-Occupied Vehicles have taken advantage of these recent advances: the *Shinkai 6500*, for example, uses rechargeable lithium-ion batteries. Lithium-ion batteries have had some problems with overheating and fire risks in some other applications, however, so some submersible operators have been cautious to adopt them. The *Alvin* submersible, for example, is still using lead-acid batteries at the moment, despite their lower energy density and longer recharge time, while designing and testing new lithium-ion battery packs to address overheating and fire risks. Ultimately, running out of battery power is what limits the duration of dives, rather than running out of life support, and lithium-ion batteries offer advantages in providing more power from less weight.

It's amazing to think how far we have come in a short time with the development of deep-diving vehicles. In 1894, Sydney Hickson published a popular science book called *The Fauna Of The Deep Sea*, in which he summarised what was known about the deep ocean at that time for a wide audience. Its preface contained a late-Victorian prediction that:

We may be able to plant the Union Jack on the summit of Mount Everest, we may drag our sledges to the South Pole, and we may, some day, be able to travel with ease and safety in the Great Sahara; but we cannot conceive that it will ever be possible for us to invent a diving-bell that will take a party of explorers to a depth of three and a half miles of water.

Yet only forty years later, Beebe and Barton reached a depth of half a mile; then just twenty-six years after that, Piccard and Walsh reached the deepest point in the ocean, nearly seven miles down; and today we have undersea craft that are far more capable than their bathyscaph when it comes to working on the ocean floor.

9

What's it like to dive in a deep-sea submersible?

Unlike an astronaut, there's not really any special training to become a scientific observer in a deep-sea submersible. Submersible pilots typically undergo a long programme of training and qualification to operate the craft, and usually have to become familiar with every part of the vehicle as engineers, ready to tackle any problems that might arise. Scientific observers have it easy in comparison: there are just a few minimum medical requirements, to ensure that sitting in a cramped and often cold position for several hours is not a problem. Beyond that, everyone diving in a submersible receives some instruction in emergency procedures. In the event of a fire or a poisonous atmosphere inside the sub, we need to know where to find and how to use emergency personal breathing apparatus – typically a mouthpiece like a scuba-diver uses, and goggles to protect our eyes from smoke or fumes. There could also be a situation where the pilot is incapacitated, so the scientific observer also needs to know what to do to bring the sub back to the surface on their own. That may involve a 'sphere release' procedure: buttons or switches that jettison the extraneous parts of the vehicle, such as manipulator arms and even batteries, allowing the section containing the personnel sphere to start to rise by its own buoyancy. And of course, we need to know how to use the communications system to talk to the surface ship in an emergency.

During an expedition at sea we're usually very busy aboard the mothership planning and preparing for a sub's dive right up until launch. Because the amount of time a sub can spend in the deep ocean is limited by its batteries, we need to make a very tight 'to-do' list for each dive, deciding exactly where we need to go and what we need to collect, measure or film. Depending on how deep we are going and which sub we are using, dives can typically last anything between four hours for a shallow dip with a short to-do list to perhaps twelve hours for a mission with a long task list at greater depths.

Once we have had a final dive planning meeting to make sure everyone from the sub team to the deck crew and the ship's bridge officers knows what they are doing, we climb aboard the sub, typically in a hangar off the aft deck at the stern of the ship. With two or three people crammed into a personnel sphere a couple of metres across for several hours, we do have to think about personal hygiene – not just making sure we've had a good wash that morning, but also that we're not wearing any products that are too smelly, such as strongly scented deodorants, perfume or aftershave. Similarly, anything that we take in the sphere with us, such as our own cameras to take pictures through the port-holes, needs to be checked to make sure it can't give off any noxious gases.

Once the hatch is shut, there's actually a moment of calm while waiting to be launched from the back deck of the ship – all the planning is over and it's too late if we've forgotten anything, so there's finally some time to reflect personally about where we're about to go and what we're about to see. The sub trundles out of the hangar on a rail system to the aft deck, where the crew hook the vehicle up to a crane or winch, typically on a A-frame that can extend out over the stern of the ship, and we start to sway as we're lifted off the deck, then the water gurgles up over the portholes as we're lowered into the ocean. Swimmers from a small inflatable boat hop aboard the top of the sub and unhook the lines from the ship – and that's another special moment, when we're no longer attached, free to head off and explore on our own. After completing final pre-dive checks, the pilot lets air out of a buoyancy tank with a high-pitched whine, and we slip slowly beneath the grip of the waves.

As we start to dive, the sunlight outside the portholes fades, and the colour of the water becomes a deeper and deeper blue, eventually turning black as we pass beyond the reach of the sun's rays. We don't actually 'drive' down into the deep ocean using our thrusters; instead, because our craft is heavier than water, we literally sink like a stone – in fact wobbling about a bit, like a pebble dropped in a pond. The sub usually creaks and groans as the pressure on the hull increases, tightening the seals where connections pass between the controls inside and equipment outside. Apart from that, the only sounds are our voices inside the sphere, the soft whirr of the fans circulating our air and the sonar signals we use to navigate and keep in touch with the ship above. Depending on the hydrodynamics of the sub's hull, we might typically sink at a rate of thirty or even forty metres per minute, taking about half an hour to drop through a kilometre of ocean.

As the sunlight fades away in the water outside, the temperature usually starts to drop as well. Temperature is the main consideration for clothing inside a sub: much of the deep ocean is cold, typically between 2 and 4 °C (35 to 39 °F) at more than two kilometres down. Some regions have warmer deep waters – the depths of the Mediterranean can be a balmy 14 °C (57 °F) – but some are even colder. When I took part in the first dives to one kilometre deep in the Antarctic, the water temperature outside the sub had dropped to *minus* 1.6 °C (29 °F) by 800 metres down, not quite freezing into ice only because of the saltiness of the water. Shallower-diving subs with acrylic hulls aren't too bad for heat loss, but the metal hulls of deeper diving subs conduct heat rapidly into the cold ocean. So it can be refrigerator-temperature when you're inside a deep-diving sub for hours on end.

Warm clothing is therefore essential for comfort, and an important safety precaution in case a sub should become stranded for a long period on the ocean floor. But in some parts of the world, contrasting warm temperatures at the surface can be a complication: in the tropics, the air temperature topsides could be 30 °C (86 °F) or more, sweltering those inside the hollow metal ball of the sub for at least an hour during pre-dive checks and launch. And subs that use an acrylic hull for their personnel sphere can become a bit like a greenhouse in the sunshine too. Wearing many layers of clothing is therefore the best option, although putting on or taking off layers can be awkward when there are several people inside a hollow ball two metres across.

Some subs have their own uniform 'dive suits' for bathynauts, which are essentially an all-in-one insulated overall, with an upper half that can be left unzipped and dangling during launch in warm waters. A woolly hat is still a good accessory, and seems to be something of a tradition: Cousteau's team famously wore red knit hats, and similar 'comforter caps' were worn before then by early hard-hat divers. Otis Barton also had a lucky hat, described by Beebe as a 'very greasy leather skull-cap',[12] which he insisted on taking with him for bathysphere dives, even delaying launch on one occasion while the crew searched for it, until Barton realised he was sitting on it.

Radio is fine for communicating with the ship while our sub is at the surface during launch and recovery, but we can't use radio once we're beneath the waves, because seawater blocks the radio frequencies that we would normally use for a conversation. During dives we therefore use sound waves instead of radio waves, via an acoustic underwater telephone system known as a Gertrude or UQC. The underwater telephone takes the acoustic signal of our voice and shifts it into a higher frequency, which is more efficient to transmit as sound waves through the water. A receiver aboard the ship or the sub then shifts that sound signal back into its original frequency, allowing an operator to listen to the original voice message. It can also send a single tone sound for Morse code, allowing routine messages such as 'everything ok' to be abbreviated to just one or two Morse code letters so our work is not interrupted.

The transmitted signal from the underwater telephone sounds like a high-pitched whistling or chirping, and you can hear it through the hull of the ship without any equipment, for example if you are off duty and lying in a cabin bunk to get some rest. You can't make sense of what's being said in that raw transmitted form, but you can tell if there is a lot of communication going on between the sub and the ship, which may be an indication that something unusual is happening. And some of the routine 'check-in' messages using Morse code do become familiar in their raw form.

Because seawater blocks the radio frequencies that are useful for communication, we can't usually transmit live video from a Human-Occupied Vehicle to the surface ship either. It is possible to encode pictures as sound signals using an 'acoustic modem' and send them to the ship, but the transmission is very slow – the picture builds up line by line over a minute or more for just one image. Remotely Operated Vehicles (ROVs) avoid this problem because they are connected to the ship by a tether that carries live high-definition video from their cameras, and recently there have been experiments connecting Human-Occupied Vehicles to ships with very thin fibre-optic cables for the same purpose. In 2013, my colleague Professor Ken Takai of the Japan Agency for Marine-Earth Science and Technology (JAMSTEC) broadcast live from the *Shinkai 6500* submersible at five kilometres deep in the Cayman Trough, via a fibre-optic cable to the ship above and then to shore via satellite, so that more than 300,000 people could join in with his dive on the internet. But thin fibre-optic cables are easily severed, and the risk of entanglement complicates dives, so we'll see whether that new type of connection becomes routine for Human-Occupied Vehicles in future.

If we're heading down to work on the ocean floor, rather than working in mid-water away from the seabed, we keep the lights switched off until we arrive, to save battery power. A sonar 'pinger' starts to tell us when we're getting close to the ocean floor, giving us a warning perhaps 100 metres above it, at which point we'll switch on the lights and the pilot will adjust our buoyancy again, slowing our descent for final approach – we really don't want to slam into the sea floor. All those aboard start to look out for the seabed appearing below, which suddenly appears in the lights. After a few more adjustments to our buoyancy so we can hover just above the seabed, it's time to start our 'to-do' list of dive tasks. There will be particular samples that we need to collect for all the different scientists involved in our expedition: perhaps rocks for geologists to analyse in their labs, sea-floor mud for geochemists and video records of the distribution and behaviour of animals for biologists, along with specimens of possible new species.

Keeping accurate records of what was measured or collected, where and when, requires a lot of concentration, but it's essential so that our colleagues can make use of the samples and data that we bring back – we're the only ones down here, working on their behalf. And when you think about the overall expedition – the cost and sometimes years of planning to get a ship and sub to this spot in the ocean – there's a lot riding on getting it right. Plus you never know when bad weather or a technical problem could make this the last dive of the expedition, so you need to make the most of the rare opportunity of being here in the deep ocean. The informal training that I received as a scientific observer, which I try to pass on to my own students, is about making careful and detailed notes and records during a dive: when and where you are, what you are doing, what you have collected, what you are seeing. Those records are essential for everyone to make use of the samples and data that the dive has collected, and I still believe in using an 'old-school' notebook and pencil for drip-proof scribbling during dives, in addition to devices such as voice recorders, just in case of any failure with that equipment.

The pressure of getting through the 'to-do' list is what preys on the mind rather than the pressure of the ocean outside, but sometimes, when we're perhaps waiting for an instrument to take measurements of something over several minutes, we get a chance to think about the astonishing place that we're visiting: the vast landscape of dark terrain stretching out for hundreds of miles in every direction beyond the pool of our lights – and we can look out of the porthole in wonder. On my first dive, I remember seeing a type of deep-sea fish called a chimaera, also known as a 'ghost shark', hanging on the other side of the port-hole, seeming to peer in with one of its large dark eyes. It struck me that we were barely two metres apart from each other, yet in completely different worlds: the fish in a world of darkness and crushing pressure, while I was in the airy and well-lit interior of our sub.

We usually take packed lunches on dives, containing whatever the ship's galley prepares, though nothing too spicy, pungent or messy. Not surprisingly, sandwiches are a staple. Chocolate and sweets are also common snacks to help keep energy up for concentration. The food may not seem exotic, but the place where we have our 'picnics' certainly is. My most memorable meal was aboard Japan's *Shinkai 6500* submersible: for our packed lunch, we had the choice of sandwiches, which were popular with Japanese colleagues as a change from their normal menu, or traditional Japanese food, which I love. So I enjoyed onigiri rice balls and a flask of green tea five kilometres deep in the ocean, looking out at hot springs on the ocean floor, while one of our instruments was making a measurement.

The packed lunches taken aboard submersibles provided an accidental experiment on microbial life in the deep ocean in 1968, when the *Alvin* submersible flooded and sank, fortunately without anyone aboard. When the sub was recovered from around 1,500 metres deep, more than ten months later, the lunches were still inside, including sandwiches made with baloney and pre-sliced white bread. They showed little signs of bacterial decay, leading microbiologists at that time to suggest that rates of microbial activity might be very slow in the chilly and high-pressure conditions of the deep ocean. This incident stimulated curiosity about microbes in the deep ocean, and further work has shown that microbes are actually very active in the deep ocean, but they are difficult to grow on Petri dishes, let alone on packed lunches.

When it comes to going to the toilet inside a submersible, it's best to try not to, as there are no facilities. Having to pee into a sealable container is possible, and a 'she-wee' funnel can assist female bathynauts. But the lack of privacy with three people crammed into a compartment two metres across encourages us not to drink much before or during a dive. As a result, some sub pilots can become susceptible to health problems caused by frequent dehydration, such as kidney stones – I've been on a couple of expeditions where the sub pilot needed treatment for that ailment, once even requiring the ship to return to shore to send them to hospital. Needing to defecate while inside a submersible isn't something to even contemplate: the smell would be pretty unbearable in the confined atmosphere, and I think the prospect would be reason enough to end a dive for the sake of all those aboard. Trying not to eat anything that could upset your digestion before or during a dive is therefore a good idea.

One of the most common challenges that we face during dives is navigation. In our everyday lives, we've become used to knowing exactly where we are in the world around us thanks to smartphones and satnavs with digital maps and signals from Global Positioning System (GPS) satellites. But what if there wasn't already a map of the area, and you didn't have any GPS signals? Imagine being dropped in the middle of some unfamiliar countryside on a moonless cloudy night, with just a torch and a compass, and having to find your way to a particular clump of trees somewhere out in the darkness. That's what we face during dives: we can't receive GPS signals from satellites while we're underwater, because seawater blocks their radio signals, so we have to use alternative approaches. Not being entirely sure where we are is an uncomfortable if all-too-familiar feeling during dives, and I am obsessive about navigation, because if we don't know where we are, then the observations, measurements or samples that we collect are much less useful.

Although we can't receive GPS signals underwater, our ship at the surface can. So if we can establish where we are in relation to the ship, we can work out the exact latitude and longitude of our underwater vehicle. One way to do this is a technique called 'ultrashort baseline' acoustic navigation, abbreviated to 'USBL'. One half of the system involves putting an acoustic beacon on our underwater vehicle, which sends sound signals up towards the ship. The other half of the system consists of several receivers lowered on a pole or retractable keel beneath the ship to get clear of its hull. Those receivers are slightly separated from each other, and by measuring tiny differences in the sound signal that each one receives, the system can work out the direction of the sound signal from the underwater vehicle.

The ship also sends out signals to the beacon on the underwater vehicle, and by measuring the time that the vehicle's beacon takes to reply, we can work out the distance between our vehicle and the ship. If we also know the depth of our vehicle, measured directly by a pressure sensor on the hull, then a bit of trigonometry can turn our distance to the ship into a 'slant range' that gives us the downwards angle from the ship to where we are in the ocean below. So from all that, we know three things: the distance from the ship to our vehicle, its direction from the ship on the compass and the angle down to it in the water – everything we need to fix the position of the underwater vehicle relative to the ship. And as we know the ship's position from GPS, we can work out the exact latitude and longitude of the underwater vehicle at the sea floor.

A computer takes care of all the maths, of course, and when the system works, it's great – but it doesn't always work perfectly. Turning the travel times of sound signals into distances depends on knowing the speed of sound through the water very accurately, and that depends on the temperature and salinity of different layers in the ocean, which can vary. Sometimes the sound signals bounce around a bit off the seabed too, particularly if the sub is diving on rugged and rocky terrain, giving echo signals that can confuse the system. The greater the depth, the greater the uncertainty, and even with a well-calibrated system, at five kilometres deep the estimated position of the sub at the sea floor might jump around by as much as fifty metres from moment to moment, which can be a problem if you are trying to navigate to one particular spot.

So to help, it is worth making a detailed map of the area of sea floor beforehand, by using an Autonomous Underwater Vehicle or a Remotely Operated Vehicle with a multibeam sonar system. If we fly that survey vehicle at an altitude of around 50 metres above the seabed, following a 'mowing the lawn' pattern over the area, we can typically map 'landmark' features as small as a couple of metres across, such as rocky outcrops. When we see one of those features during a dive at the sea floor, we know where we are on the map, even if the acoustic navigation system is jumping around a bit in its estimates of our position.

To help guide us to underwater 'landmarks' during dives, we can also use 'sector-scanning' sonar from our underwater vehicle, which sends a beam of sound sweeping out around us, rather like the beam of light from a lighthouse. Sonar can 'see' much further out into the water than we can with our lights, so it can tell us more about our surroundings, for example if we are near any underwater cliffs or mineral spires poking up from the sea floor. From that, we can usually figure out where we are on our detailed map, and where we need to go.

At the start of my career, as part of my PhD research I had to make a map of the mineral spires of some undersea hot springs on the Mid-Atlantic Ridge, newly explored by the two *Mir* submersibles of the Russian Academy of Sciences in the mid-1990s. Those subs were using an older and more complicated system of underwater navigation, involving a network of beacons at the sea floor, and because of the rugged underwater terrain, most of the time the subs couldn't determine their position from those sea-floor beacons. So to make my map from their dive videos afterwards, I had to resort to old-fashioned dead-reckoning, estimating distance travelled and compass direction when the subs moved across the sea floor, and using landmarks on the ocean floor to tell when they were visiting the same places at different times. It took me a few months to piece together a map using that crude method, instead of a couple of days at most that it would have taken if the navigation system had worked okay. But when a US expedition returned to the area a few years later with the *Alvin* submersible and a more successful set-up for its navigation, I was encouraged to see how my map of sea-floor features broadly matched the one that they made.

Since then, I have kept the habit of old-fashioned navigating-by-landmarks during dives, just in case there is a problem with the automated system. And it has paid off. In 2010, I took part in an expedition exploring undersea hot springs near Antarctica with a Remotely Operated Vehicle. During our early dives, there were some teething troubles with our USBL navigation system, but I had sketched a map in my head – and then on paper – of the features that we encountered and my guesstimates of their relative locations, which was sometimes enough to put us back on the right track. The only downside was occasionally being woken up in my bunk while off-watch, to come to the control centre and help navigate to targets when the vehicle got lost.

The clock is always ticking down on our dive time in a Human-Occupied Vehicle, and sooner than seems possible it's time to leave and head back to the sunlit world far above. Adjusting our buoyancy again allows us to start to float back up to the surface, through the long dark of the ocean's interior. Very gradually the light from above becomes apparent – we perhaps notice it because of the equipment on the outside of the sub starting to become visible again through the portholes. Once we're back at the surface, we bob about for a while as the ship manoeuvres to pick us up. If the seas are choppy at all, we'll be rolling rather than bobbing during this wait, and this is when people get seasick – it's not pleasant inside the cramped sphere, but fortunately at least it's the end of the dive. Swimmers hook the sub up to the crane aboard the ship, then water streams off the portholes as we're hauled out, and we arrive back on deck with a clunk.

There's a final wait for a few moments while we equalise the pressure inside the sphere – although the sphere keeps us at normal atmospheric pressure, our breathing always produces a slight difference in pressure compared with the outside world by the time we get back. Then the hatch opens and we climb out, welcoming the chance to stretch our limbs again and – hopefully – feel sunlight on our faces, perhaps followed by a fairly urgent visit to the toilet. If it's someone's first dive, there may be a ritual of dumping a few buckets of seawater over them to baptise them as a bathynaut. Then we help our colleagues unload the 'sample basket' on the front of the sub and make sense of what we collected, measured and recorded during the dive. Sorting through the samples from the dive, and processing them as required to preserve or analyse them, can take several hours, and we usually need to write up a formal dive report as well while the memories and details are still fresh in our minds.

Eventually we can flop into our bunks, usually exhausted but exhilarated after a good dive. And meanwhile, members of the sub team are checking the vehicle's systems, charging the batteries, replacing the chemicals in the CO_2 scrubber and possibly changing oxygen cylinders, to get ready for the next dive.

What are the dangers of diving in submersibles, and why don't we just use Remotely Operated Vehicles instead?

Human-Occupied Vehicles are engineered and tested to withstand conditions beyond their maximum operating depth, so pressure crushing the hull is not usually a risk. But other dangers during dives include the sub becoming trapped at the sea floor, or crashing into it, and electrical fires from equipment aboard. And although Human-Occupied Vehicles have made thousands of successful dives into the deep with a very low rate of accidents overall, unfortunately there have been some.

On 17 June 1973 the *Johnson Sea Link* submersible became entangled in debris at a depth of 110 metres, while collecting a fish trap from a shipwreck at 9:46 a.m. local time. The *Johnson Sea Link* carried people in two separate chambers: two people in a transparent acrylic sphere at the front of the vehicle, and two more people in a separate aluminium chamber behind, designed to allow its occupants to make 'lock-out' dives, leaving the vehicle to work on the sea floor. While they were stuck on the sea floor, the temperature in the aluminium rear chamber became much colder than in the acrylic sphere, because the metal walls conducted heat more rapidly. Carbon dioxide then built up in the atmosphere of the rear chamber, because the chemical scrubber system usually absorbing it from the exhalations of the occupants became less effective in the cold conditions.

In the rear chamber, Albert Stover and Clayton Link – who was the son of the submersible's designer Edwin Link – put on personal breathing equipment as the CO_2 scrubber failed. The pressure inside the rear chamber began inexplicably to increase, and Stover and Link were forced to switch to breathing a helium-oxygen mix as the pressure climbed to that of the depth outside the sub, twelve times normal atmospheric pressure. But breathing a helium mix chilled their bodies further, as the gas conducts heat more effectively than air, making them unable to attempt a risky 'lock-out' dive to escape the chamber. By 12:30 a.m. on 18 June, Stover and Link were no longer communicating with pilot Jock Menzies and marine biologist Dr Robert Meek in the forward chamber, and at 1:12 a.m. Menzies reported to the surface that the two men in the rear compartment were suffering convulsions.

A grappling line eventually snagged the sub and it was hauled to the surface at 4:53 p.m., after thirty-one hours trapped at the sea floor, but there were no signs of life in the rear chamber. The medical examination of Stover and Link's bodies recorded their approximate time of death as 1:00 a.m., and 'respiratory acidosis due to carbon dioxide poisoning' as the cause. The investigation of the accident recommended that Human-Occupied Vehicles should thereafter be equipped with a life-support system that remains effective for at least seventy-two hours in case the occupants need rescue from the deep. The *Johnson Sea Link* sub returned to service, modified to reach 1,000 metres deep, and was a superb vehicle for science, but I found it impossible to avoid thinking about what Stover and Link went through when I dived in its rear chamber years later.

There have been several other incidents involving Human-Occupied Vehicles that were not fatal, but which illustrate the risks of sending people into the deep. In August 1973, for example, the *Pisces III* submersible was working off the coast of Ireland, laying a new transatlantic telephone cable. As the sub was about to be lifted out of the water onto its mothership after a dive, a towing line wrapped around one of the vehicle's engineering compartments, tearing it open and allowing seawater to flood in. The sub therefore lost buoyancy and started to sink, with crew members Roger Mallinson and Roger Chapman still aboard. The sub eventually hit the sea floor at 480 metres deep, travelling at around forty miles per hour, but remained intact.

The crew estimated that they had sixty-six hours of breathable air supply, as Mallinson had fortunately recharged the tanks just before their last dive, and the pair started to do whatever they could to conserve it. Two other *Pisces* submersibles were rushed to the scene, but their initial attempts to find and attach lifting lines to the stricken sub were not successful. Eventually a Remotely Operated Vehicle managed to attach lines to the sub, and *Pisces III* and its occupants were rescued, seventy-six hours – three days – after plummeting to the sea floor. There were just twelve minutes of air supply remaining when the hatch was finally opened. The *Pisces III* incident remains the deepest sub rescue in history. Chapman went on to design the *LR-5* rescue sub, originally operated by the Royal Navy, to bring crews home from sunken submarines.

There were also incidents during the first dives by Human-Occupied Vehicles to the Mid-Atlantic Ridge in 1973 and 1974. On the second dive to the ridge in 1973, the bathyscaph *Archimède* experienced a dip in its electrical power supply while at the sea floor, which automatically triggered the release of its ballast to return to the surface. Not too dramatic; but then on the way up, an electrical fire somewhere behind an instrument panel quickly filled the personnel sphere with acrid smoke, forcing the occupants to don personal breathing gear. Aboard the bathyscaph that day was Dr Bob Ballard, already a veteran of using the *Alvin* submersible for science, and whose subsequent career as an ocean explorer would include taking part in the discovery of hydrothermal vents and finding the wrecks of the *Titanic* and the *Bismarck*.

In 1974, the *Alvin* submersible joined in the exploration of the Mid-Atlantic Ridge, and on 17 July set off for the ocean floor with pilot Jack Donnelly and scientists Bill Bryan and Jim Moore aboard. During the dive, they encountered a fissure on the sea floor wide enough for *Alvin* to enter, which would enable them to see geological features not visible from above, and to look for possible volcanic heating of the seawater inside. But as the sub proceeded forwards and downwards along the fissure, the walls narrowed around and above it. Attempts to manoeuvre out by moving up, forwards or back went nowhere. Fortunately, the scientists' careful notes describing their entry into the fissure, and their observation of drifting particles of detritus to deduce the currents at the sea floor, allowed Donnelly eventually to retrace the sub's movements and escape.

In 1977, Ballard was again aboard a bathyscaph, this time the *Trieste II* in the Cayman Trough of the Caribbean Sea. On this occasion the bathyscaph met the rugged sea floor sooner than expected, without the usual time to slow its descent carefully by dropping ballast, and the bow of the buoyancy tank crashed into the volcanic terrain. Through a porthole, Ballard spotted petrol leaking from it, and the pilot immediately ditched all the remaining ballast weights to start the ascent to the surface. But with petrol bleeding out, the bathyscaph was losing buoyancy. The ascent took six hours, and the occupants could do little but hope that enough buoyancy would remain to return them to the world above. Fortunately, it did.

On a lighter note, bathynauts have also had to consider the risks of possible attacks by marine life, though these have proved to be less of an issue. In 1948, Cousteau's team prepared the *FNRS-2* bathyscaph with equipment that could fend off an attack by a giant squid, or attempt to snag a specimen for science, in the form of an underwater gun firing an electrified harpoon. But that untested equipment had to be removed from the bathyscaph for its eventual dives, and giant squid have proved to be very elusive to bathynauts, rather than aggressive. But on 6 July 1967, the *Alvin* submersible was rammed by a swordfish while investigating deep-sea coral off the coast of Florida. There was no damage to the personnel sphere, but the sword of the fish became stuck between two parts of the outer casing of the sub, just below one of the portholes. Those inside aborted the dive, and the stuck fish was later served as barbecue. *Alvin* had a similar scrap with a marlin a couple of years later, and these incidents raised concerns about the risks of a porthole itself being rammed, if fishy aggression was perhaps triggered by seeing light inside. But experiments simulating the impact of a swordfish swimming at full speed confirmed that a porthole could withstand such an assault.

I'm glad to say that I have had hardly any scary moments during dives in submersibles. On my first dive in 1995, aboard DSV *Sea Cliff*, the US Navy pilot stood up once we reached the ocean floor at 2,200 metres deep and unbolted the hatch at the top of the sphere. This was apparently an initiation ritual for scientists on dives, presumably to see whether we might freak out at that point. The hatch is a thick wedge of metal, and the pressure of the ocean forces it shut against the hull. No human would be strong enough to actually push it open against that pressure, so it is safe to undo the mechanism holding it shut while at depth. Realising that, I managed to keep calm, apart from the overwhelming excitement of being at the ocean floor for the first time.

Some things happen on dives that might seem scary, but are actually fairly routine. The penetrators that carry electrical connections through the hull, for example to control external equipment such as the sub's mechanical arms, sometimes have seals that may leak a bit, particularly at shallower depths where the pressure outside isn't yet pushing them tight. Higher pressure from going deeper can settle the seals, cutting off any dribbles. The same can be true for the portholes on some deep-diving subs with metal hulls: higher pressure actually pushes the cone-shaped windows in more tightly against the hull, stopping an occasional bit of seawater from weeping in at shallow depths. So when you find drips or even a puddle of water inside a sub, the usual procedure is to dip a finger in and taste it. If it's fresh water, then it's just condensation, and a quick dab with a towel takes care of it. But if it's salty, it is worth tracing the source and keeping an eye on it, but not immediately assuming the worst – minor leaks often resolve themselves.

My scariest moment was when we started to smell burning plastic during a dive, which became very unpleasant very quickly in the confined atmosphere inside the sub. An electrical fire is probably my greatest fear, and in some subs we deliberately lower the oxygen content of the air down to about 16 per cent to reduce fire risk without impairing those inside – but not in this one. The pilot halted our descent, and we quickly unplugged all the additional equipment that we had taken into the sub for filming on that dive, such as cameras and lights with their power supplies. The next option would have been to put on personal breathing gear and goggles and abort the dive quickly, but fortunately whatever was burning stopped when we unplugged everything, and as the air cleared we were able to continue.

Human-Occupied Vehicles (HOVs) and Remotely Operated Vehicles (ROVs) are both 'human-directed' vehicles, ideal for the 'investigate the anomalies' step of ocean exploration. But there is sometimes heated debate about whether one type is better than the other for that task. Remotely Operated Vehicles offer several advantages in addition to not posing any risk to human life. Although ROVs still need pressure-resisting compartments to keep their electronics dry, those compartments can be much smaller than the chamber that carries people into the abyss, so they are cheaper to build than Human-Occupied Vehicles. And because ROVs must be connected to a ship by a tether, so that their operators aboard ship can see the live video from the cameras at the sea floor and control the vehicle, they can also be powered via their tether cable, unlike free-swimming HOVs that need to carry batteries. With no people or batteries aboard, ROVs can be smaller and more manoeuvrable than HOVs, and also work far longer at the sea floor for each dive, with a team of operators aboard ship controlling the vehicle around the clock in shifts. This advantage becomes increasingly important in deeper water, where the proportion of time spent travelling to and from the sea floor is greater in the time-limited dives of Human-Occupied Vehicles compared with longer dives by ROVs.

For example, in February 2013 I led an expedition to the Cayman Trough, where we spent 45 per cent of our eighteen-day expedition actually working on the sea floor with our ROV – not bad given that we had to bring it back from time to time to unload the samples that it collected, carry out occasional maintenance of the vehicle and also complete some tasks with the ship's other scientific equipment. In contrast, the best that a Human-Occupied Vehicle would have been able to achieve, diving every day of a hypothetically similar expedition, would have been around 25 per cent of expedition time working at the sea floor. Most HOVs usually dive for a maximum of twelve hours per day, because of the need to recharge their batteries between dives, and at the depth we were working, an HOV would also have spent about six hours of each day's dive just travelling to and from the sea floor. So our ROV enabled us to spend nearly twice as long working at the sea floor as a Human-Occupied Vehicle would have done.

Furthermore, instead of just carrying one scientist inside the sub on a dive, during ROV dives an entire scientific team aboard ship can direct activities, interpreting discoveries and making decisions together. On my expeditions, I like to ensure there is a biologist, geologist and a representative of any other relevant type of scientist in the team for each shift in the control centre. With modern high-bandwidth satellite communication between ship and shore, scientists elsewhere in the world can now be involved in ROV dives too, watching live sea-floor video over the internet and chatting with operators aboard ship. This 'telepres-ence' capability is now routinely used by the Ocean Exploration

Program of the US National Oceanic and Atmospheric Administration from its ship the RV *Okeanos Explorer*, and by Bob Ballard's Ocean Exploration Trust from the EV *Nautilus*. It's a surreal experience to sit in comfort ashore watching the ocean floor live from an ROV and helping to direct its operations with colleagues around the world, but it shows how far we have come in the eight decades since Beebe and Barton first peered through the portholes of their bathysphere.

Some science ROVs also have a precision-flying mode known as 'closed-loop control', which makes them far more manoeuvrable than current Human-Occupied Vehicles. In closed-loop control, the ROV bounces a sound signal off the sea floor beneath it, and measures any change in the frequency of the sound that comes back. Any 'doppler shift' in the sound – the kind of difference in pitch that we hear between the siren of an ambulance coming towards us and when it's moving away from us – tells an on-board computer how the vehicle is moving relative to the sea floor below. The computer can then automatically activate the necessary thrusters to keep the vehicle steady in a fixed position, adjusting those thrusters all the time if the vehicle is buffeted by ocean currents. It's amazing to see a Remotely Operated Vehicle the size of a family car keeping itself still like that, more than two kilometres below the control centre aboard our ship.

Once a Remotely Operated Vehicle is in closed-loop control, we can drive it extremely precisely, programming movements into its computer rather than piloting it with a joystick. For example, we can tell the vehicle to hover above one spot on the sea floor, keep facing the same direction and just move vertically upwards three metres. Then we can tell it to stay at the same depth and move right by just half a metre, and then move vertically downwards three metres, all the time facing exactly the same direction. Using this method, we can 'fly' the vehicle in a precise vertical plane facing an underwater feature such as the mineral spire of a hot spring on the ocean floor, all the time recording close-up video of that feature using a forward-looking camera. If we then take slightly overlapping video frames from along that flight path and stitch them together, we can build up a complete image of the whole underwater feature, effectively 'scanning' it with our forward-facing camera.

This technique, refined by my research team, can give us a complete view of an underwater feature that would be too large to see in a single camera image, because by the time we moved our vehicle back far enough to see it all, it would have faded into the darkness beyond the reach of our lights. And because the overall picture of the feature is built up from hundreds of individual images, we can zoom in on fine details, even counting the numbers of limpets living on the backs of crabs in the overall image. So we can use the technique to recreate a whole underwater landscape digitally, which then allows us to explore its patterns of life just like a biologist surveying a rocky shore to find out where different species live on it. And it's all thanks to the precision control capabilities of Remotely Operated Vehicles; Human-Occupied Vehicles don't currently have closed-loop control systems, and their greater bulk can make them more susceptible to drifting in ocean currents.

Nevertheless, I still think there is something special about immersing yourself in the environment that you are trying to understand, and the exceptionally wide-angle and three-dimensional view human eyes give us from a Human-Occupied Vehicle provides an intimate perspective that is somehow different to that from the cameras of a Remotely Operated Vehicle. On my first dive in a Human-Occupied Vehicle, what struck me most as I looked through the portholes was the dramatic patchiness of life around the undersea hot springs that we were investigating. That perspective has provided the focus for some of my research since then, trying to understand a pattern that could be grasped instantly from actually being there, but which can be much harder to perceive from a Remotely Operated Vehicle without some fancy flying and lots of image processing afterwards. And Human-Occupied Vehicles may well catch up with some of the current capabilities of the Remotely Operated Vehicles, for example with thin fibre-optic tethers to share video with the ship above, and perhaps, in future, closed-loop control. I am glad that we have the two different types of vehicles available, and I think there is plenty of room for both of them in exploring the deep ocean.

Is the deep ocean divided into 'zones' at different depths?

Not really. There are often schematics of depth zones in older textbooks and on some websites, but they're not really meaningful, and perhaps hark back to a time when we used to think of 'the deep ocean' as a single environment. Those schemes typically include a 'bathyal zone', an 'abyssal zone' and finally a 'hadal' zone, at increasing depths. But let's consider the 'bathyal zone', which usually refers to 200 metres to 4,000 metres deep (though sometimes it's given as 200 to 2,000 metres, or even 1,000 to 4,000 metres, and that lack of consistency should perhaps alert us that it might not be that useful a notion).

A 'bathyal zone' with a depth range of 200 to 4,000 metres includes a variety of different habitats such as continental slopes with their occasional undersea canyons, much of the 60,000-kilometre mid-ocean ridge including the sea-floor hot springs dotted along it, some shallower parts of the muddy abyssal plains and many rocky underwater mountains. About the only feature shared among them is pressure conditions of 20 to 400 times normal atmospheric pressure, as a result of the depth range defining that 'bathyal zone', and those pressure conditions are not the most important feature governing what lives there or the processes that take place in those different environments.

If we were to adopt a similar scheme on land, lumping everything between 200 and 4,000 metres above sea level into a single 'zone', it would include contrasting habitats such as rainforests, the Atacama Desert, much of the Antarctic Peninsula, most of the Great Plains of the United States, the Highlands of Scotland, the South Downs of England and a lot more. It would therefore be pretty impossible to make any useful generalisations about the physical conditions of that 'zone', or the biology of its inhabitants. So zoning schemes for the ocean based solely on depth don't reflect the variety of habitats and conditions that we now recognise in the deep ocean.

Instead, I think it's more useful to divide the deep ocean according to environments with different conditions. In the mid-water of the deep ocean, away from the sea floor, there are two zones that we can define in that way. We first came across them both in the Question 6 chapter: first there's the dysphotic zone, also nicknamed the 'twilight' zone. Here there is still some sunlight filtering down from the surface but it's too faint for drifting microscopic algae to thrive. Then there's the deeper aphotic zone, also nicknamed the 'midnight' zone, which is beyond the reach of any photons of light from the sun.

If we dive out in the very clearest waters of the open ocean, aboard an undersea craft launched on a bright morning from a research ship far from land, then the first 200 metres or so of our plunge can be relatively well-lit by sunlight, before we reach the twilight of the dysphotic zone. This uppermost layer, beneath its canopy of quicksilver, is home to drifting microscopic algae known as phytoplankton. These tiny yet numerous inhabitants use sunlight energy to grow through the process of photosynthesis, similar to plants on land. And like plants on land, their photosynthesis turns carbon dioxide into sugars, and releases oxygen: in fact, around half the oxygen in our atmosphere – or every second breath we take – comes from those dwellers of the sunlit upper ocean.

As we look out of our porthole, the sunlight gradually fades as its rays are quenched by the seawater and scattered by its infinitesimal inhabitants, and by 200 metres deep or possibly earlier, we have entered the dysphotic or twilight zone of the deep ocean.

The remnants of the sun's rays are the key feature of the dysphotic zone, because everything that lives in the faint downwelling light potentially casts a shadow beneath it. Predators and prey therefore play hide-and-seek in the twilight, trying to spot the shadow of their next meal above while trying to conceal their own shadows from would-be diners below. Ninety per cent of the species in the dysphotic zone are bioluminescent, producing light displays from organs on their bodies, often to help with their hide-and-seek.

The upper limit of the twilight zone is usually defined at 200 metres deep, and the boundary between the twilight and midnight zones is usually given as 1,000 metres deep. But the depth boundaries between these two zones vary in different regions of the ocean: sunlight doesn't penetrate as far into the ocean where there are abundant tiny floating algae that make the water near the surface more turbid. Looking out of the porthole of a mini-sub returning from a deep dive in the clear waters of the Caribbean, I could start to make out a white plastic tray on the front of the sub at a depth of 427 metres with my not-too-sensitive human eyes. But during a dive in the Antarctic during summer, when microscopic algae were blooming at the surface and making the water very murky, it looked pitch-black to my eyes outside the sub at a depth of eighty metres. The boundaries of the mid-water zones vary in time too: every evening when the sun sets, the upper boundary of the 'twilight' zone rises all the way up to the surface, because at night there isn't enough light for algae to carry out photosynthesis. And with less light shining down from the surface, the upper boundary of the 'midnight' zone becomes shallower at night as well.

The dysphotic zone is the stage for the greatest animal migration on Earth, much larger in biomass than all the herds of wildebeest lumbering across the African savannah. Every evening ten million tonnes or so of marine life rises slowly from the inky depths to feed overnight in shallower waters, before sinking back down at dawn. There is more food available nearer the surface where microscopic algae thrive, but feeding there during the daytime involves a greater risk of being eaten by predators that hunt using their eyesight. Hence the commute to spend the night in shallower waters, with the animals moving up and down to stay at the same level of light as dusk falls and dawn breaks – it's actually the upper boundary of the dysphotic zone that moves as the sun sets and rises, and its inhabitants move with it. Marine biologists first became aware of this phenomenon by watching the traces of echosounders, which show a dense marine life – nicknamed the 'deep scattering layer' because sonar echoes off it almost like the sea floor – rising towards the surface as night falls. And when the sky darkens during solar eclipses, echosounders also show that marine life beginning its commute just as if it were dusk.

When diving in a submersible, the view outside often looks like stars streaking upwards past you, but that impression is created by you sinking past the small animals living in the water, and overtaking the particles of 'marine snow' – mostly formed from the faeces of tiny animals in the water – that are sinking more slowly to the bottom. But sometimes that 'starfield' becomes noticeably thicker for a few minutes, when you're passing through a deep scattering layer of more abundant mid-water life. By one kilometre deep, the last vestiges of the sun's rays have faded away into the water, and we have passed into the aphotic or midnight zone, beyond the furthest reach of sunlight. But it is not utterly dark: many of the species that live here still have bioluminescence for hunting, confusing or deterring predators and for attracting mates, and many of them still have eyes. And if you flash a light from a submersible out into the darkness, some of the bioluminescent animals around the sub flash back at you a second or so later. Light is the language that they use to communicate, and a flash of light starts a conversation with some of them – although at the moment we don't really know what our flash is saying to them, or what their reply means.

The mid-water aphotic zone is our planet's 'inner space', and its vast volume provides most of the habitat for life in the oceans. In 1971 an aircraft encountered a Rüppell's vulture soaring at an altitude of eleven kilometres above the coast of west Africa, which might tempt us to think that the atmosphere could actually offer a more voluminous habitat for animal life than the oceans. But animals that take to the air don't live their entire lives up there – at some point, they land somewhere to, for example, lay their eggs. In contrast, many ocean animals spend their whole lives drifting or swimming in the ocean without ever encountering a hard surface, and consequently many of them have soft gelatinous bodies that look almost ethereal, like blown glass. As a result, animal life in this realm is perhaps the least known on Earth, because early attempts to collect specimens in nets resulted in their delicate bodies disintegrating, leaving just unrecognisable 'snot' in the net. Today we can observe those animals in their natural habitat from our underwater vehicles, but global databases of observations of marine species show that the deep mid-water world is chronically under-studied, even compared with the deep sea floor.

Some of the conditions at the ocean floor also vary with depth, but are not consistent at the same depth in different places, so we can't really use them to define 'zones' for sea-floor environments. Water temperature varies with depth, usually getting colder as we go deeper because cold water is more dense and therefore sinks. But it's not a gradual change: rather, there are layers of water with different temperatures and salinity at different depths. So as a sea-floor feature such as a continental slope drops down into greater depths, its surface can be bathed by different layers of water along it. Those layers may have different marine life associated with each of them, leading to 'bands' of different species living on the sea floor at different depths, sometimes with abrupt changes between them as we move across the boundary into another layer of water.

But the layers of water vary in different regions of the ocean. Along the Reykjanes Ridge, which runs south-west from the coast of Iceland down to a depth of more than three kilometres, the upper 1,000 metres or so of the ridge crest is bathed in a layer of water called 'subpolar mode water', which can have a temperature of 6 °C. The layer beneath that, covering deeper parts of the ridge, is called 'north Atlantic deep water', and it can be less than 4 °C. And if we were to move into deeper water off the flanks of the ridge, we could find an even colder layer of 'Antarctic bottom water'. In contrast, the Cayman Trough of the Caribbean Sea only has a single layer of water below 1,200 metres deep, all the way to the bottom more than five kilometres down, and that layer has a monotonously consistent temperature of around 4.5 °C.

The chemistry of the oceans also varies with depth around the world, with consequences for the types of sediment that settle at the sea floor. The temperature, pressure and amount of dissolved carbon dioxide in seawater can combine into conditions that help to dissolve calcium carbonate beyond a particular depth in the ocean, known as the carbonate compensation depth. Calcium carbonate forms the tiny coatings of some of the plankton that live in surface waters, and any particles of 'marine snow' made from their remains will start to dissolve if they sink beyond the carbonate compensation depth. So the sea floor deeper than that depth is covered by particles made of other materials, such as the remains of organisms that have coatings of silica instead of calcium carbonate. But there's also a deeper point where silica starts to dissolve in the ocean, called the opal compensation depth, and the only sediments that can settle beneath that are particles of clay, blown from land by the wind.

The depths where carbonate and silica shells and skeletons dissolve away in marine snow vary around the world: for example, the carbonate compensation depth in some parts of the Atlantic is around five kilometres, compared with 4.2 kilometres deep in some parts of the Pacific. And what settles at the sea floor also depends on what starts at the surface – if a region has very few plankton with siliceous shells and skeletons living at the surface, then the sea floor below will not be dominated by that type of sediment, regardless of its depth. So the ocean floor is covered in different types of sediment like a patchwork quilt, and depth alone does not predict its pattern. Instead of dividing the sea floor into depth zones, then, I prefer to consider categories of sea-floor environments formed by different processes, such as mid-ocean ridges, abyssal plains, seamounts, ocean trenches and continental slopes – and those broad categories each contain a further variety of different types of habitats for us to explore.

12

What are the strangest places you've been to on the ocean floor?

There's lots to choose from, because the deep ocean floor seems so alien to us, so I'll pick three places. The first is the world's deepest known undersea hot springs, five kilometres (3.1 miles) down in the Cayman Trough of the Caribbean Sea. These undersea hot springs, known as 'hydrothermal vents', are like the geysers of Iceland or Yellowstone, but they erupt continuously on the ocean floor. The water that gushes out of them is hotter than molten lead – these ones in the Cayman Trough are 401 °C – but it doesn't turn into steam because of the pressure of all the water above.

The hot water is also rich in dissolved minerals, which form tiny particles when mixed with cold seawater at the bottom of the ocean. This is what gives deep-sea vents their nickname of 'black smokers', because those particles make the jet of hot water look like a column of dark smoke. The minerals in the hot water also solidify to build tall spires rising up from the seabed, nicknamed 'chimneys' because of the hot, smoky-looking water billowing out of the tops of them. These vent chimneys can be as slender as your leg, but as tall as a two- or three-storey house, looking like twisting columns of solidified smoke.

Some of the vent chimneys that we found at the bottom of the Cayman Trough were particularly slender and beautiful, and yet the jet of hot water erupting from them rose more than half a mile into the ocean above. Tucked in beside them in our submersible, I was aware of the awesome force of nature right next to us – I could almost feel the volcanic fury of the hot fluid roiling out of the chimneys, as if they were the outlets of some furnace in the Underworld.

But alongside these scalding vent chimneys there's a lush garden of deep-sea life: there were swarms of pale and orange-tinted shrimp crowding their slender sides, and the sea floor between them was carpeted with speckled white anemones with translucent tentacles. Bizarrely, these animals are ultimately sustained by the mineral-rich fluids gushing from the vents, which nourish bacteria that in turn provide food for the animals. It's definitely the most impressive as well as the strangest place that I've visited; this is where the old-fashioned elements of fire and water combine, perhaps to forge life itself.

On my dive there, which was the first time a mini-submarine had visited those particular hot springs, our first few hours went really well: we were blitzing through our 'to-do' list, and I was starting to feel a bit smug about that. But when manoeuvring the sub for our next task, we got caught in the jet of water shooting out of the mineral spires and carried up away from the sea floor by it. By the time we got out of it and dropped back down, we weren't in the same place – and we ended up spending much of our remaining dive time trying to find our way back, in the end not quite completing our task list. The deep ocean can teach us lessons in hubris as well as wonder.

My second choice of strangest place is a 'brine pool' at the bottom of the northern Gulf of Mexico. There are lots of brine pools scattered about on the sea floor in this area; this one is roughly oval in shape, about twenty metres long and fifteen metres wide and named Brine Pool NR-1 after the US Navy's *NR-1* submarine that mapped it in detail. In contrast to the vigorously roiling hot springs of the Cayman Trough, the brine pool is an eerily calm place, and much shallower at 650 metres deep on the continental slope. The ocean currents are weak at the sea floor here, so the pool doesn't mix into the ocean above but lies like a lake at the bottom of the sea, nestled in a dip at the top of a gentle hill of soft mud. It has a visible surface that we can even park our submersible on, as if we were a boat floating on a lake. The water in the pool sometimes has a slightly milky look as the lights from our sub catch it at certain angles, and sometimes it looks like there are tendrils of mist rising from it too – a spooky scene in the surrounding darkness.

Beneath the muddy sediments that form the seabed here, there's a layer of salt, created when this part of the sea was cut off from the rest of the ocean and dried up about 170 million years ago. The ocean has since returned, so now there is a stack of salt, sedimentary deposits and finally the ocean itself above. The salt at the bottom of the stack is more buoyant than the sediment on top of it, so it pushes its way upwards, eventually seeping out of the sea floor as the brine that fills the pool.

The water of the pool itself can be deadly to deep-sea animals, because there is little or no oxygen dissolved in it. From the *Johnson Sea Link* submersible, I watched eel-like fish occasionally blunder into the pool and immediately writhe back in shock, often but not always managing to escape. But around the edge of the deadly pool, which looks like the shore of the lake, there's a five-metre-wide bed of brown-shelled mussels with other animals crawling among them, including orange-striped squat lobsters with long slender claws, pale shrimp, snails with shells that look a bit like overgrown toenails and 'snotworms' that wind their slimy bodies into knots. And towards the outer edge of the mussel bed, where their shells become sparse, there are some yellowish tubes sticking up out of the mud, each a bit wider than a pen and some reaching waist-high, with the feathery red plumes of tubeworms just showing out of their trumpet-shaped tops. This is another lush garden of life, richer than the surrounding sea floor: a 'cold seep', here nourished by fluids being squeezed out of the seabed rather than by hot mineral-rich water shooting out of a hydro-thermal vent as in the Cayman Trough.

As the salt beneath the sea floor rises up through the sediments, it also pushes out hydrocarbons and methane from the breakdown of organic matter buried in the sediments. Oil naturally oozes out of the sea floor in a few places here; looking over the side of a ship you can sometimes even see blobs of oil arriving at the sea surface like bubbles. The methane that also seeps out of the seabed nourishes microbes that in turn provide food for some animals – the mussels that surround the brine pool have methane-fuelled microbes living inside their gills, for example. Other animals around the brine pool graze on bacteria that grow in mats on the sea floor or on the mussel shells, and some animals prey on those grazers too. Consequently there's a rich colony of deep-sea life here, fuelled by what's seeping out of the seabed with the brine.

Some microbes buried in the sediment feed on the seeping methane and produce hydrogen sulfide as a by-product, so this chemical, which gives off a 'rotten eggs' smell, also seeps up out of the seabed here. The hydrogen sulfide sustains the red-plumed tubeworms at the edge of the mussel bed: their tubes grow down into the sediment, and the worm inside the tube taps into the hydrogen sulfide seeping up from below. The worm has a bag-like organ inside its body, packed with bacteria that use the hydrogen sulfide as an energy source, and those bacteria in turn provide food for the worm.

As the microbes living in the sea floor feed on the seeping methane, they also create carbonate rock as another by-product. Eventually, in a few centuries' time, that growing deposit of rock just beneath the muddy seabed will block off the rising brine and methane here, and the present inhabitants of the brine pool will die out. But the rock built by the microbes may poke out of the seabed, providing a rare island of rocky habitat among the surrounding soft mud, and end up colonised by deep-sea corals that need to attach to a hard surface to grow. So the occasional thickets of deep-sea corals dotting the sea floor away from the brine pool may once have been 'cold seeps' just like this one, and a part of a centuries-long cycle of life that we have only just begun to glimpse.

*   *   *

My final choice of strangest place is another deep-sea garden, this time in the chilly waters of the Antarctic, where life is lavish for yet another reason. The continental shelf around Antarctica is deeper than the shelves around other landmasses, reaching 500 to 600 metres deep, because the weight of the ice sheet on the land pushes down the continent. Close in to shore there are also sea-floor gullies more than a kilometre deep, scoured out by glaciers in the past, when the ice sheet was more extensive. So although the continent of Antarctica is remote, requiring research ships to cross the often stormy waters of the Southern Ocean to reach it, once you get there the deep sea is right on the doorstep. And although conditions are harsh for life on land, they are rich for life beneath the waves, at least in summer. Twenty-four-hour daylight allows tiny drifting algae, phytoplankton, to photosynthesise round the clock and thereby bloom in the sunlit surface waters. Those algae provide food in particular for krill, shrimp-like crustaceans about as long as a little finger, which feast on the algae during summer.

Krill are an important part of the food chain for marine life in the Antarctic, providing sustenance for populations of several species of penguins and whales that breed there in summer. But I hadn't really appreciated just how important krill are for life in the deep, too, until I took part in the first dives by Human-Occupied Vehicles to reach one kilometre deep in the Antarctic. Cousteau had taken his 400-metre-rated 'diving saucer' to the Antarctic in the early 1970s, but his expedition was beset by problems; and since then only a few dives by Remotely Operated Vehicles had explored the deep Antarctic sea floor. So I jumped at the chance to be 'scientific guide' for dives there in 2016 with the BBC Natural History Unit, who mounted an ambitious expedition to film in the deep ocean there for their *Blue Planet II* television series.

At the tip of the Antarctic Peninsula, which is the finger-like projection of the continent that points up towards South America, there's a stretch of water called Antarctic Sound between the mainland and a couple of offshore islands. That narrow seaway provided a haven for our dives in the acrylic-hulled *Deep Rover 2* submersible, protected from the wind and waves in most directions, although we did have to dodge occasional icebergs drifting along the channel.

The water was greenish-brown at the surface with all the algae blooming in it, and their thick soup made it dark by just eighty metres deep. I could then see a blizzard of tiny lights from krill around the sub – they have eight lamp-like bioluminescence organs in two rows down their bodies – and switching on our own lights attracted them to the sub in a dense swarm, right up against the see-through hull. But I could also see string-like things drifting in the water among them, just a millimetre or so thick and longer than the krill themselves.

These string-like things were initially a mystery, until I saw a krill squirting one out of the narrow 'nozzle' of its posterior: we were diving through krill poo, which was sinking softly to the sea floor below.

Krill have a particular behaviour that helps their poo to fertilise the deep: after they have gorged on plankton near the surface, they enter a state of torpor, rather like us needing a snooze after a heavy meal. During their post-prandial nap, the krill stop swimming and sink down to deeper waters, and when they rouse themselves afterwards, they usually relieve themselves too, releasing their poo closer to the sea floor than where they fed. That means there's less chance for other animals to make a meal of their droppings before they arrive at the sea floor; so the defecatory behaviour of krill provides a fast-food delivery service to the deep sea floor. As a result, that sea floor is blanketed by animals that catch a meal from the poo that falls like snow and others that plough through it after it has settled.

On the seabed of the Antarctic Sound we found yellow feather-stars with arms like dusters, which took off if they sensed the bow-wave of the sub nearby, dancing through the water by waving their arms in a mesmerising pattern before sinking back down like parachutes to settle on a new spot. There were sponges as big as barrels, with mantis-like crustaceans crawling on them and waving their legs up into the water to filter food from it. Darting among them were dragon-faced icefish, whose clear blood lacks the red haemoglobin pigment that carries oxygen in our own blood. The cold temperature of the deep here means that more oxygen dissolves directly into the fluid of their blood, so the icefish can manage without red blood cells. There were also huge sea spiders with leg-spans forty centimetres across, striding over the sea floor like the long-legged alien machines from *War Of The Worlds*, and sometimes our lights would silhouette a grainy-skinned octopus gliding about with its arms trailing behind it like a cape.

Overall it's an astonishing living landscape, with krill poo supporting the richest life I have ever seen at one kilometre deep, away from colonies of creatures clustered at hydrothermal vents and cold seeps. And together those three environments offer a snapshot of the many ways that life thrives in the dark of our deep-ocean planet.

What are the most exciting discoveries
so far in the deep ocean?

There are two discoveries in particular that stand out for me: one that changed our understanding of how our planet works, and another that rewrote the rules of biology. The first discovery led to the second, and it all began with the tenacity of an ocean explorer kept ashore by the sexism of the decades in which she lived.

Marie Tharp joined the Lamont Geological Laboratory of Columbia University in 1948, after the Second World War had opened up higher education in science and technology for women in the US. But women were not permitted on research voyages, so Tharp worked ashore, collating echosounder traces recorded by ships at sea. By piecing together detailed echosounder depth measurements from ships crossing the Atlantic, she began to see a narrow valley consistently running along the crest of the Mid-Atlantic Ridge – something that looked very much like a rift valley on land, where the Earth's crust splits apart and magma wells up from below.

The idea that such a process could be happening on the ocean floor, and taking place all along the middle of the Atlantic, was scientific heresy at the time. Alfred Wegener had proposed the theory of 'continental drift' back in 1913, noting similarities between the coastlines of Africa and America to suggest they were once joined and then rifted apart, but his idea was still considered 'fringe science' at best. Tharp's boss and long-standing collaborator Bruce Heezen was therefore reluctant to accept her evidence for a rift valley running along the Mid-Atlantic Ridge, and literally sent her back to the drawing board to redraft her map. Tharp persisted, however, and in 1952 Heezen also plotted the epicentres of subsea earthquakes detected in the Atlantic, finding that their locations matched up exactly with Tharp's valley, as would be expected if the sea floor was rifting apart there.

Jacques Cousteau was also originally sceptical of the undersea rift idea, but in 1959 his team lowered a camera to the deep Mid-Atlantic Ridge for the first time, and took photographs of a rift valley on the ocean floor there. Tharp and Heezen also traced the mid-ocean ridge beyond the Atlantic, showing that it snakes all around our planet, like the seam on a tennis ball. And evidence that emerged in the late 1950s and early 1960s for 'sea-floor spreading' and 'plate tectonics' – the movement of the plates of the Earth's crust – supported the idea that the sea floor was rifting apart along that vast ridge.

The realisation that the mid-ocean ridge was a 'sea-floor spreading centre', where the plates of the Earth's crust were rifting apart and new sea floor being formed by volcanic activity, was boosted by studies of the magnetic properties of sea-floor rocks on either side of the ridge. Rocks formed from lava at the sea floor carry a record of the Earth's magnetic field from the moment at which they solidify. Every few million years, the Earth's magnetic field 'reverses' when north and south magnetic poles swap over. So rocks formed at the same time have the same magnetic signature recorded inside them, and the rocks of the sea floor on either side of the mid-ocean ridge form matching 'stripes' of identical magnetic signals, parallel to the ridge. That peculiar pattern, confirmed by Fred Vine and Drummond Matthews in the early 1960s, only makes sense if the rocks in matching stripes on either side of the ridge were actually formed together at the same time on the ridge, and have since rifted apart. That evidence helped to cement the final acceptance of plate tectonics, which revolutionised geology as profoundly as revealing the structure of DNA revolutionised biology.

*   *   *

We now recognise the 60,000-kilometre mid-ocean ridge as our planet's greatest geological feature – as Tharp remarked later, 'you can't find anything bigger than that, at least on this planet'[13] – and it was the massive missing piece in the puzzle of how our planet works. The plates of the Earth's solid crust ride on convection currents in the fluid mantle beneath them, and the mid-ocean ridge lies where those plates are being pulled apart, with molten rock extruding to create new sea floor and fill the otherwise growing gap between them. As a result of this process along the Mid-Atlantic Ridge, Europe and North America are gradually moving away from each other, slightly more slowly than the rate at which our fingernails grow.

The mid-ocean ridge is the edge of creation, where new crust of the Earth's surface is forged, and as the new plate is pulled away from the ridge it gradually gets buried by sediments sinking down from the ocean above. Close to the ridge, the sediments have not yet built up, and the terrain of the sea floor can be dominated by knobbly abyssal hills that have been produced by sporadic lava eruptions at the ridge. But further from the ridge, the sediments have had more time to build up as the plate moves away from the ridge, forming a layer up to a kilometre thick and burying the rocky terrain of the crust beneath a muddy abyssal plain. Finally, there is an opposite edge of destruction where the plate eventually collides with another to form an ocean trench, recycling the crust back into the furnace of the Earth's interior. Realising that this dynamic process inexorably resurfaces our planet was a unifying 'Rosetta Stone' moment for geology, enabling earth scientists to make sense of observations in areas ranging from palaeontology to geophysics.

To explore the mid-ocean ridge close up, French geophysicist Xavier Le Pinhon wrote to US geologist Ken Emery in 1971 with the idea of bringing together the deep-sea tools of both nations for an investigation of the Mid-Atlantic Ridge south of the Azores. This led to the creation of Project FAMOUS, which stood for French-American Mid-Ocean Undersea Study, and in 1973 the bathyscaph *Archimède* carried people to the deeply submerged mid-ocean ridge for the first time, four years after the first astronauts walked on the Moon. The smaller and more nimble *Alvin* and *Cyana* submersibles joined in during 1974, and Project FAMOUS obtained photographs and rock samples that transformed researchers' understanding of the vast geological processes shaping the sea floor. Scientists saw the rocky rift with their own eyes through the portholes of those craft, and wrote afterwards in the journal *Science* that 'the ocean floor is disturbingly different from what we had imagined'.[14]

But the mid-ocean ridge was the epicentre of yet another scientific revelation, a decade after its recognition as the geological backbone of our planet. In February 1977, the Galapagos Hydrothermal Expedition was using two ships – the RV *Knorr* and the RV *Lulu* – to investigate an area of mid-ocean ridge near the Galapagos Islands, where previous expeditions had detected unusually warm water near the seabed, and even photographed a pile of large white clamshells on the usually barren lava of the ridge. Through the night of 15 February into 16 February, the team, including Bob Ballard, aboard the RV *Knorr* used a towed camera to photograph more patches of white clam and brown mussel shells, in the same area where their temperature probe measured warmer-than-usual water near the seabed at a depth of 2.5 kilometres. Launching from the RV *Lulu* on 17 February, the 713th dive of *Alvin* targeted that area, carrying geologists Jack Corliss and Tjeerd van Andel along with pilot Jack Donnelly. Their expectation was to find hydrothermal vents, and they did indeed see shimmering and cloudy water rising from cracks in the lava of the sea floor. The water had this shimmering appearance because it was at a warmer temperature than the seawater around it, like the heat haze rising above a road on a hot day; and the cloudiness was caused by particles of minerals forming in it as it mixed with cold seawater. The temperature probe aboard DSV *Alvin* recorded 8 °C, balmy in comparison with the surrounding chilly temperature of the deep sea.

Corliss, van Andel and Donnelly became the first people to visit a deep-sea hydrothermal vent, but what they did not expect to find was a riot of deep-sea life thriving there: clams with shells up to thirty cm long, scattered across fifty metres of sea floor, along with beds of brown-shelled mussels, numerous white crabs and a purple octopus. 'Isn't the deep ocean supposed to be like a desert?' Corliss called up to the RV *Lulu* from *Alvin* via sound-powered telephone. 'Well, there's all these animals down here.' Subsequent *Alvin* dives also found a patch of benthic siphonophores – gelatinous animals that look a bit like dandelions – and a field of giant tubeworms, with bright red plumes poking out of white tubes up to half a metre long. But there were no biologists aboard the RV *Knorr* or RV *Lulu*, as the goal of the expedition was to study the geology and geophysics of the mid-ocean ridge, building on the work of Project FAMOUS in the Atlantic. So the scientists preserved specimens of these mysteriously abundant deep-sea animals as best they could in vodka, for biologist colleagues to study ashore.

The puzzle for biologists was how so many animals could be thriving at 2.5 kilometres deep, where food reaching them by sinking from above is rare because most of it gets eaten on the way down. The answer lay in other samples collected by the expedition team: water collected from the vents stank with the 'rotten eggs' smell of hydrogen sulfide. That sulfide provides energy for microbes, which in turn provides food for the animals at the vents – a new kind of food chain discovered in the deep ocean, supported by a process known as chemosynthesis.

Chemosynthesis had been predicted as a possibility by microbiologists in the nineteenth century, and been found in a few exotic microbes elsewhere, but it had never before been found supporting a food chain of animal life. In chemosynthesis, microbes use energy released by the oxidation of chemicals in the fluids gushing out of the seabed, instead of using sunlight energy for the process of photosynthesis that generally nourishes plants and algae. Discovering whole colonies of creatures fuelled by chemosynthesis at hydrothermal vents opened the minds of scientists to new ways that life can thrive beyond the reach of sunlight. It also gave us new ideas about how life on Earth might have begun around four billion years ago, and eventually raised the possibility of life in the oceans beneath the icy crusts of some of the moons of Jupiter and Saturn. Reporter David Perlman of the *San Francisco Chronicle* was aboard the RV *Knorr*, and he wrote exclusive accounts of the expedition's discoveries for the newspaper's readers, predicting in a 9 March article that 'When these findings are all analyzed in detail they are bound to "revolutionize" many theories about the deep ocean floor', beneath a headline telling of 'Astounding Undersea Discoveries', neither of which were media hyperbole.[15]

In 1979 *Alvin* returned to the mid-ocean ridge near the Galapagos, this time with biologists and a *National Geographic* film crew aboard, to study the life around the vents in more detail. The sub and its mothership then headed north to another section of mid-ocean ridge, called the East Pacific Rise, where French scientists had been investigating the sea floor. On 21 April, geologists Bill Normark and Thierry Juteau were aboard the sub with pilot Dudley Foster, at latitude 21 °N on the East Pacific Rise. They came across a mineral spire, two metres tall, with a jet of smoke-like black fluid gushing out of its top. Attempts to measure the temperature of the smoke-like fluid maxed out the 91 °C limit of *Alvin*'s temperature probe, and in fact destroyed its plastic tip, which had a melting temperature of 180 °C. Later dives with a different probe measured a temperature of more than 350 °C at the source of the fluids gushing from this 'black smoker' vent, with the pressure at the sea floor preventing those fluids from boiling into steam. These scalding 'black smoker' vents were therefore unlike the balmy warm vents near the Galapagos, but thanks to chemosynthesis were still home to abundant deep-sea life. The very high temperatures of black smoker vents are only found right in the throats of their 'chimneys', and water temperatures typically drop by more than a hundred degrees within a few centimetres away, so the animals around them live at cooler temperatures, similar to those at the Galapagos vents.

The smoke-like appearance of the vents comes from minerals dissolved in the hot fluids until they mix with cold seawater to form metal-rich particles. Understanding the source of the dissolved minerals in the vent fluids, and their impact on the ocean, has helped geochemists to understand what governs the chemistry of seawater. The metal-rich precipitates from the fluids also build the spire-like 'chimneys' of the vents, and seeing that process has given geologists insights into the formation of ore bodies mined on land – and raised the possibility of mining such metals on the ocean floor.

Since these discoveries of the late 1970s, we have found hydrothermal vents on the mid-ocean ridge all around the world, and they remain a focus of investigations for geologists, geochemists and biologists trying to understand our deep-ocean world. I have spent most of my career exploring them, in the Atlantic, Pacific, Indian and Southern Oceans; there are also some at the bottom of the Arctic Ocean, but I haven't managed to get there yet.

14

What are the weirdest creatures you've seen in the deep ocean, and what are your favourite new species that you've discovered?

The weirdest creature that I've seen is probably a benthic siphonophore, related to the ones mentioned in the previous question and known as a 'flying spaghetti monster'. I saw one at 650 metres deep in the Gulf of Mexico. I love it because it looks like a creature from the early series of *Doctor Who* that I watched on TV when I was growing up. It has a 'body' like a shaggy barrel, about a metre tall, and a slender 'neck' that it can extend and retract like an old-fashioned radio aerial. At the top of that 'neck' is a structure that works like a float, for buoyancy, and looks a bit like a 'head'. Then there are also a few long thin tentacles that dangle from the body. The whole weird-looking creature drifts along in the gentle currents near the seabed, seeming to float across the terrain like a ghost, trawling the sea floor behind it with its long tentacles. When it encounters a rock, it raises up the float structure on its extendable neck, then pulls its body up after it to sail over the potential hurdle.

It's actually not a single animal, but a colony of polyps, with different types doing different jobs. Those that form the trailing tentacles catch prey, others form the float structure and some have responsibility for reproduction. They all share a common digestive system, so that prey caught by the hunters can be shared with the whole colony, and all the polyps are linked by a communal nervous system to coordinate the colony. The Portuguese Man O' War that lives at the surface of the ocean is the same type of animal, but its cousins that live at the deep-sea floor look even more bizarre. And there are other siphonophores that drift in the interior of the ocean, with chains of polyps that hang in the darkness like a string of fairy lights and can reach forty metres long, which is longer than a blue whale.

Worm-like animals called 'enteropneusts' are another of the weirdest life forms I've seen in the deep. The ones that crawl over the ocean floor belong to a family of species whose Latin name translates as 'neck-plough' worms, and their bodies are divided into a bulbous and often-colourful 'head' and skinnier translucent 'tail', reaching up to a metre long overall. Rather than living in burrows like many worm-type animals, they creep across the sea floor at a rate of about ten centimetres per hour thanks to tiny hair-like structures, called cilia, on the underside of their bodies. As they go, they sweep their 'heads' across the surface of the sea floor – hence the name 'neck-plough' worms – sucking in sediments to feed on and leaving a continuous trail of faeces behind them. Their faeces provide a useful marker of their passage: some lay a line that wanders back and forth, while others leave a trail that winds round in a spiral. They were given the nickname 'spiral-shit animals' by deep-sea biologists who saw those patterns of poo in sea-floor photographs. Once they have exhausted a patch of food on the sea floor, they can lighten their bodies by emptying their bowels further and drift up into the water, sometimes oozing mucus to make a balloon around their bodies, and get wafted on gentle currents to a new patch of sea floor.

One species, *Yoda purpurata*, gets the 'Yoda' part of its scientific name from the bulbous parts of its head that look like the ears of Yoda from the *Star Wars* films, and the 'purpurata' part of its name describes its purple colour. Other species have heads that range in colour from orange to dark blue, so enteropneusts are examples of how some deep-sea animals can still be colourful, thanks to the colours of whatever their bodies are made of, rather than all white or colourless down there in the dark. And enteropneusts also demonstrate how much we still have to learn about life in the deep ocean: so far, only seven different species of deep-sea 'neck-plough worms' have been described, but there are undoubtedly more out there. One species has even been found carrying the embryos of its offspring around on its body, each wrapped up in a membrane and nestling in a tiny dip on its mother's back.

A final example of the weird and wonderful that I am very fond of is the 'Hoff' crab: a scarab-like eyeless crustacean whose undersides are coated in hairs rather like a carpet-cleaner. Or, as my former PhD student Dr Nicolai Roterman observed when we first collected a specimen, like the famously hairy chest of actor David Hasselhoff, hence the 'Hoff' crab nickname. The crabs are about three to ten centimetres long from the back of their tucked-in tails to the tips of their claws, and their carapace often looks off-white to yellowish in colour. The scientific name of the crab is *Kiwa tyleri* – the second part of that name honours the career of deep-sea biologist Professor Paul Tyler – and the species lives around hydrothermal vents at 2.4 and 2.6 kilometres deep near Antarctica. They are related to the 'yeti' crabs with long hairy arms that live around hydrothermal vents in the equatorial Pacific, as Nicolai showed in his PhD research.

I first saw Hoff crabs in 2009 while on an expedition towing a camera beneath the RRS *James Clark Ross* to find hydrothermal vents on the ocean floor. But it was only a glimpse of them on that reconnaissance mission, as we didn't want to risk damaging the hydrothermal vents with a camera swinging around on its wire while the ship rolled in the swell of the Southern Ocean. That glimpse was initially puzzling, because crabs and other crab-like animals are not supposed to survive in the cold deep waters of the Antarctic. In cold conditions, crabs and their relatives cannot get rid of magnesium that builds up in their blood from seawater, and it acts as a narcotic, stupefying them to the point that they can no longer feed. But we theorised, mindful of the monkeys that bask in warm volcanic pools amid the winter snows of Nagano Prefecture in Japan, that the crabs we had glimpsed could be living in warm waters around the hydrothermal vents to avoid that chilly fate.

In 2010 we returned with a Remotely Operated Vehicle aboard the RRS *James Cook*, which allowed us to map out where the Hoff crabs were living around the hydrothermal vents, and collect specimens to find out what they eat and how they reproduce. Both of these qualify as weird: the Hoff crabs feed on bacteria growing on the hairs of their chests, combing them off to eat them. The bacteria are nourished by the life-giving waters of the hydrothermal vents, and the crabs pile together in their thousands with warm mineral-rich fluids trickling up among them to ensure a good crop from their chest-hair farms. But those piles of crabs are a poor place for females to rear their offspring, which they carry as embryos tucked beneath their tails until they hatch. So the males and females mingle in the piles, and then the females crawl further away from the hydrothermal vents to brood their embryos. Bacteria no longer grow on the chests of the crabs in the cold water away from the vents, so females cannot feed while they wait for their embryos to hatch. But afterwards they crawl back to the vents, running a gauntlet of predators, to mingle with the males again. The males, meanwhile, remain in the warm fluids where they can continue to feed, and consequently they can grow much larger than the females. So male and female Hoff crabs live quite different private lives, to balance the demands of feeding and reproducing. My former student Dr Leigh Marsh figured this out as part of her PhD research.

One of the volumes of Pliny the Elder's epic *Naturalis Historia*, which he wrote at the end of his life between 77 and 79 CE, lists 'the names of all the animals that exist in the sea, one hundred and seventy-six in number', with the bold assertion that 'by Hercules! in the sea and in the ocean, vast as it is, there exists nothing that is unknown to us'.[16] Fortunately for marine biologists, Pliny was rather premature in declaring 'mission accomplished' for exploring life in the ocean depths. By analysing the rate at which we are finding whole new categories of animals – new families, orders and classes of organisms – and looking at how many species there usually are in those categories, scientists estimate than only around 10 per cent of species living in the oceans have so far been described and given scientific names.

To describe a new species and give it a scientific name, we have to show how it is different from other species that we already know. This involves detailed examination of its anatomy, to find features that distinguish it from other known species and which can be used to identify it in future. We usually also compare parts of its genetic code with those of other species, to understand how the new species is related to them. When other scientists have checked that work to make sure we haven't made any errors, we publish the formal description in a scientific journal. This includes giving the new species its scientific name.

There are some rules about what names can be given to new species – you can't name a species after yourself, for example, but you can name it after a person who is not involved in describing it, for example to commemorate another scientist. But traditionally the scientific name is devised to help distinguish the species, for example describing a key feature that defines it, or the region where it was found. So we might call a species 'aculeata' if it has sharp spines, because that name comes from the Latin word for sting, or a species discovered in the Scotia Sea of the Antarctic might be named 'scotiaensis' to indicate where it lives.

One of the new species that I have been involved in describing with my research team is a shrimp that lives at hydrothermal vents in the Cayman Trough of the Caribbean Sea, which we named *Rimicaris hybisae*. My former student Dr Verity Nye led the description of this new species, which was one of many that she described during her PhD research. The shrimp are about as long as a little finger, and thousands upon thousands of them swarm on the sides of mineral spires of the hydrothermal vents in the Cayman Trough – there is one colony at 2.3 kilometres deep there, and another at five kilometres deep. Unlike most other shrimp, the adults of this species don't have eyes on stalks; instead, they have a light-sensitive patch on their backs that can detect the glow that comes from hydrothermal vents, which is too faint for us to see.

The shrimp feed on bacteria that grow on their mouthparts, and they have an appendage with a brush-like tip to sweep them up to eat. The bacteria use the mineral-rich fluids of the hydrothermal vents as an energy source, so the shrimp are often found jostling around the vents to find the best spot for them to flourish. *Rimicaris hybisae* has some close relatives that live at other hydrothermal vents elsewhere, including *Rimicaris exoculata* at vents on the Mid-Atlantic Ridge and *Rimicaris kairei* at vents in the Indian Ocean, but the Cayman species differs from those in the shape of its light-sensing organ and some of its other features.

Rimicaris hybisae is one of my favourite new species because of the effort that went into collecting the first specimens to analyse. During an expedition in 2010 when we first saw the shrimp, we were working with an underwater vehicle called *HyBIS*, which stands for Hydraulic Benthic Interactive Sampler – this was not a fully functional Remotely Operated Vehicle but basically a manoeuvrable grab, lowered beneath the ship on a wire, with a camera attached. (The species name 'hybisae' celebrates the success of the *HyBIS* vehicle in discovering it.)

The grab was unlikely to be able to collect any shrimp, because they would swim out of its grasp. What we really needed was a suction sampler on a Remotely Operated Vehicle or a Human-Occupied Vehicle, but neither were aboard as the expedition was scouting for hydrothermal vents, with plans to return later with an ROV if we found any. So with the engineers aboard, we cobbled together an *ad hoc* suction sampler to attach to *HyBIS* instead, using a couple of plastic buckets, a tube normally used to collect samples of sediment from the sea floor and an underwater pump stripped from a water sampling system. Unlike a proper suction sampler on a Remotely Operated Vehicle, we could not control the pump directly from the ship to switch the suction on and off, and had to rig a timer for it instead.

We set the timer to give us a couple of hours to get the vehicle five kilometres down the hydrothermal vents where the shrimp were living but, almost inevitably, we got lost when we reached the sea floor, desperately hunting for the mineral spires where we had seen the shrimp before. We found them with minutes to spare before the timer hit zero, and then watched as the pump came on and sucked a few shrimp up the tube into the sample chamber that we had fashioned from the buckets. Our sampler was a precarious lash-up, but it worked. The few shrimp that it collected became the 'type specimens' that defined the new species, and they are now archived in the Natural History Museum in London so that future scientists can compare them with other shrimp.

Close to where *Rimicaris hybisae* shrimp congregate around some of the hydrothermal vents in the Cayman Trough, there are also eel-like fish basking in the warm waters seeping from the sea floor, with purplish-grey bodies about ten centimetres long, dark eyes and prominent lips that give them a languid-looking expression. They belong to another of my favourite new species, which we named *Pachycara caribbaeum*. When I led an expedition to the Cayman Trough with a Remotely Operated Vehicle in February 2013, we carefully collected the first three specimens of this previously undescribed species, using the ROV's suction sampler. Working with colleagues at the Natural History Museum in London, we then analysed the internal anatomy of one of them using a CT scanner, similar to the ones used to examine patients in hospitals. Those scans revealed what the fish had just eaten (the outlines of *Rimicaris hybisae* were visible in its stomach) and eventually enabled my undergraduate student Russell Somerville to show how its skeleton was different from those of other fish species. But it was what happened while we were getting ready to prepare the formal description of the new species that makes *Pachycara caribbaeum* one of my favourites.

In October 2014 my former research students Dr Leigh Marsh and Dr Diva Amon were aboard the EV *Nautilus* of the Ocean Exploration Trust, exploring the deep sea on the other side of the Caribbean, about 2,000 kilometres from the Cayman Trough. Their expedition was live-streaming video from its ROV on the internet, and I was able to follow along from the comfort of my sofa at home in the evenings. The expedition discovered an amazing outcrop of methane hydrate – a bizarre type of methane ice that can form under the seabed and occasionally pokes out of it – just over one kilometre deep, near the island of Tobago. Methane bubbles out of the ice-like hydrate when it is exposed at the sea floor, and that methane provides energy for microbes, which in turn support a lush colony of animals at the ocean floor, like the colonies at hydrothermal vents. So there were spectacular beds of brown-shelled mussels here, similar to those at the cold seeps of brine pools, with shrimp and crabs crawling among them. And as the ROV's camera panned across the scene in close-up, I could see some eel-like fish that looked like those at the hydrothermal vents in the Cayman Trough, although the ones here were much smaller.

The ROV collected one of the fish, which we then analysed to show that it was a juvenile of the same species. That meant we had found a new species whose distribution spanned 2,000 kilometres across the Caribbean, which posed lots of still-unanswered questions, such as: do the juveniles live at shallower methane seeps, and the adults live at deeper hydrothermal vents? Is the population near Tobago separate from the one in the Cayman Trough, or do fish migrate between them, making journeys similar in scale to those undertaken by eels travelling between the ocean and rivers? This often happens in ocean exploration – one discovery prompts lots of new questions as we become aware of something in the deep that we didn't know about before – and that's why *Pachycara caribbaeum* is one of my favourites.

Describing a new species involves a branch of science called taxonomy, which uses a hierarchy to classify organisms, showing how they are related to each other. We can think of that hierarchy as being like a tree, with different species as its thinnest branches. Several thin branches can be attached to the same slightly thicker branch, which is attached in turn to an even thicker stem branch, and so on. If species are the thinnest branches, then the slightly thicker branch to which they are attached is the next category up in the hierarchy of organisms, known as a genus. And then genera (the plural of genus) belong to a family, then an order, then a class and finally a phylum, which we can imagine as the trunk of the tree.

In the deep ocean, we don't just find new species of animals – we also discover thicker branches in the tree of life that no one has seen before, representing new genera, families and even orders of organisms. I've been involved in describing a new family of sea stars – also known as starfish – that have seven arms, unlike the familiar five-armed ones that live in rock pools on our shores. One of the species in the new family of sea stars lives around hydrothermal vents in the Antarctic, where its diet includes female Hoff crabs that crawl away from the vents to brood their offspring. Finding a new family in the hierarchy of life means that you've discovered an organism that is very different from others known so far, and shows how much life there is still to explore in the oceans, contrary to what Pliny reported nearly two thousand years ago.

15

How do animals survive the conditions in the deep ocean?

For a long time, scientists didn't think that life was possible in the conditions of the deep ocean. In the early 1840s, the naturalist Edward Forbes dredged up animals from down to 420 metres deep in the Aegean Sea, and noted from his hauls a decline in the abundance of life with depth. From those observations, he extrapolated that the deep ocean might become devoid of animal life beyond 550 metres deep. Unfortunately for Forbes, the Aegean Sea is unusual in having particularly sparse deep-sea life on its continental slopes because of its circulation, and it is not representative of most of the deep ocean. And Forbes was seemingly unaware of some earlier discoveries of life from greater depths, starting in the mid-eighteenth century when fishermen in the Caribbean, using very long lines, snagged strange animals resembling palm fronds from the deep sea floor. They sent those specimens to naturalists in Europe, who identified them as stalked crinoids – animals well known from fossils, but previously thought to be extinct in today's oceans – and in 1767 Carl Linnaeus named them as a new species, *Isis asteria*. They were the first deep-sea animals to be collected and described, though the exact depth from which they came was uncertain.

In the early nineteenth century, the French naturalist Antoine Risso also described several new species of shrimp and fish collected by fishermen on long-lines dangling one kilometre deep in the ocean near Nice, and in 1818 an expedition led by John Ross searching for the Northwest Passage snagged a sea-floor-dwelling basket star on a sounding line at 1.6 kilometres deep. In the late 1860s, the voyages of HMS *Porcupine* and HMS *Lightning* collected specimens from more than four kilometres deep in the north-east Atlantic, and then the voyage of HMS *Challenger* in the 1870s laid notions of a lifeless abyss firmly to rest by collecting specimens of marine life from all depths around the world.

Forbes's hypothesis was disproved; but that's how science advances, testing predictions based on previous data with new data, and his work stimulated interest in exploring the limits of life in the ocean. Although the *Challenger* expedition revealed the ubiquity of life at the sea floor, however, the idea of an uninhabited abyss echoed on for a few more decades in views of life in mid-water, between the surface and the seabed. Expedition member Henry Moseley, delivering a Friday evening lecture at the Royal Institution in London on 5 March 1880, suggested that 'It is quite possible that a vast stretch of water between the surface and the bottom is nearly or absolutely without life.'[17]

We now know that is not the case, and that the ocean depths are inhabited from the surface to the bottom of the deepest trenches, and away from the sea floor as well as upon it and within it. As air-breathing land-dwellers, to us the ocean depths perhaps seem 'extreme' because we cannot survive the conditions down there. But the conditions in which we live, with the weak pressure of a thin gaseous atmosphere, regular exposure to bright sunlight and so on, would seem hostile to a deep-sea animal. Hypothetical aquatic citizens of Atlantis might similarly postulate that life above the waves, beyond the shores lapped by tides, must be impossible.

So how do animals survive the conditions in the deep ocean? All animal life needs oxygen to survive, and most of the deep ocean has plenty of oxygen in its waters to sustain animal life. That oxygen reaches the deep ocean from the atmosphere via currents that start at the poles. Although in the polar regions the surface of the sea freezes in winter, not all of the seawater does; the process leaves behind very salty cold water that sinks because it is heavier than normal seawater. Surface waters in the polar regions can also be chilled by winds to form this sinking water. As water sinks into the deep ocean, it carries oxygen that has dissolved into it from the atmosphere, and the deep oxygen-rich water formed in polar regions then spills out into ocean basins around the world. The water at the bottom of the Mariana Trench in the Pacific, for example, came from the Antarctic and contains oxygen that dissolved from the atmosphere there.

Some areas of the deep ocean have more oxygen than others, depending on how far they are from where their water left the surface, because marine life uses up the oxygen as it flows through the deep. There are also some areas known as 'oxygen minimum zones', where the amount of oxygen in the water is particularly low, even to the point of preventing most animals from living there. These zones can form where there are very large amounts of detritus sinking from above and being broken down by microbes, which use up oxygen in the process. But overall, life survives in the deep thanks to oxygen flowing from the polar regions, where the deep oceans 'breathe in', which is one of the reasons why any environmental changes taking place at the poles are important for our whole planet.

Pressure may seem like the greatest challenge for life to contend with in the deep ocean, but its effects on marine life are perhaps not what we might think. When we build a Human-Occupied Vehicle to take people into the deep ocean, we need to engineer a thick hull to withstand the difference in pressure between the outside and inside of the craft. But unlike a Human-Occupied Vehicle, the bodies of most sea-dwelling animals don't have air-filled spaces inside them. Their bodies consist of solid tissues and liquids, and those materials are not compressible like air. If you want to explore that for yourself, take a plastic syringe of air, put your thumb over its nozzle and push its plunger: the air inside the syringe is easy to squeeze into a smaller volume. But try to do the same with a syringe of water: you won't be able to squeeze the liquid inside into a smaller volume. And, of course, the same would be true if you filled a syringe with something solid like putty.

Most sea-dwelling animals don't have to withstand the pressure of the deep ocean pushing in on their bodies, because there are no compressible materials inside them, unlike our air-filled deep-diving vehicles. The pressure inside the body of an animal such as a shrimp at the ocean floor is the same as the pressure outside its body. Consequently, most deep-sea animals do not 'explode' when we bring a specimen up to the surface – that's probably the question I get asked most often about deep-sea animals, and some people seem disappointed by the answer! Because there are no compressible materials in the bodies of most marine animals, nothing expands inside them as they come up; their bodies of solid tissue and liquids remain unchanged. If you partially fill a syringe with water and cover its nozzle with your thumb, then try to pull the plunger further out, the liquid inside won't expand.

Pressure does affect the inhabitants of the deep sea at a molecular level, however. Living cells depend on the enzymes that control chemical reactions inside them, and enzymes are large protein molecules. Cells are perpetually building those protein molecules, by stringing together smaller molecules called amino acids. The amino acids form chains, like beads on a necklace, and then the real magic of life happens: the chain of amino acids spontaneously springs into the complex three-dimensional structure of protein, as if some beads were attracted to each other by magnets inside them. That process is called 'protein folding', and the final three-dimensional shape of the protein depends on the types and order of amino acids in the original chain. The shape of the final protein is critical for it to do its job as an enzyme: if the protein is slightly misshapen, it won't work as well or possibly even at all, potentially stalling whichever chemical reaction it is supposed to control in the cell. As enzymes control all of the living processes inside a cell, the effects of a misshapen enzyme can be widespread and even lethal.

The high-pressure conditions of the deep ocean can trap water molecules on the chains of amino acids being built in cells, preventing them from folding into the right shape to work as enzymes. But life in the deep ocean has evolved several solutions to this. Some deep-sea species use slightly different sequences of amino acids than life elsewhere to build their enzymes, and their deep-sea recipes are more forgiving when it comes to proteins folding into the shape they need to work as enzymes. Many deep-sea animals also have 'chaperone' molecules that help their proteins to fold up into the right shapes. 'Chaperones' are small molecules that pull the water off the unfolded chains of amino acids, allowing the protein then to fold up correctly. Fish and some types of crustaceans use a small molecule called trimethylamine oxide – or TMAO for short – as their chaperone for protein-folding. The deeper a fish or crustacean species lives, the more TMAO it has in its cells, to counteract the effects of greater pressure.

But TMAO does have some 'side effects' on the amount of water stored in cells, and those side effects may be what limits the maximum depth at which fish can live in the ocean. The deepest fish seen so far are snailfish at 8,178 metres in the Mariana Trench, filmed by a baited lander experiment in May 2017. Measurements of the increasing amount of TMAO in the tissues of fish with depth predict that its side effects would become problematic only slightly deeper than that, so the bottom of the deepest trenches may well be fish-free, despite Jacques Piccard and Don Walsh reporting that they saw a flatfish at Challenger Deep from the bathyscaph *Trieste*.

Fish with cartilage skeletons, such as sharks and rays, have more TMAO in their bodies at the surface than fish with bone skeletons, because they also have to protect their proteins from another molecule called urea that controls the amount of water in their bodies. So as the amount of TMAO in their cells increases with depth, they run into problems with its side effects far sooner than do bone-skeleton fish. This may be why sharks appear to be limited to depths shallower than four kilometres, and are not lurking in ocean trenches – the deepest known shark is the Portuguese dogfish (*Centroscymnus coelolepis*), found at 3,675 metres. But some of the invertebrates that range down to the bottom of the deepest trenches, such as sea cucumbers, have evolved other molecules than TMAO to keep their proteins in shape, and they don't run into the same limitations as fish.

16

How deep do whales dive, and can they get 'the bends' like scuba-divers?

As air-breathing animals, whales face some of the same challenges as we do when it comes to diving deep with unprotected bodies. But some whales can dive nearly as far as the deepest sharks can swim, despite having to hold their breath to do so. The deepest dive by a whale wearing a temporary tag to record its depth is 2,992 metres, by a Cuvier's beaked whale (*Ziphius cavirostris*). Beaked whale species have relatively small bodies and routinely dive deep to feed on prey such as squid; a colleague of mine once said they are really best thought of as deep-living animals that occasionally come up to the surface for air, rather than air-breathing animals that sometimes dive deep.

Sperm whales may reach similar depths of around three kilometres on some dives, based on studies of the contents of their stomachs, though none have been yet been recorded diving that deep while wearing monitoring tags. And there are tantalising hints that some whales might go even deeper, to just over four kilometres deep. My former research student Dr Leigh Marsh and her colleagues recently analysed sea-floor pictures from around four kilometres deep in the eastern Pacific, and found marks on the seabed that match those sometimes made by whales at shallower depths. Some of the anatomical features of beaked whales suggest that they could potentially withstand conditions at five kilometres deep, and hopefully researchers will be able to analyse sediments from the marks on the sea floor to see if they contain any traces of whale DNA left behind by their creators.

We might think that animals that dive by holding their breath, such as whales, would not be susceptible to decompression sickness because they are not breathing in gas under high pressure like a human diver with a scuba cylinder. But if the air they breathe in at normal atmospheric pressure at the surface becomes compressed as their lungs are squeezed during a dive, then in theory they could encounter problems similar to scuba divers breathing air under high pressure. Nitrogen dissolved into their blood from their pressurised lungs could form bubbles if they ascend very rapidly, and those bubbles could block small blood vessels, damaging tissues and organs. Human breath-hold divers have learned to contend with this problem, for example adopting procedures in freediving competitions such as extended rest periods between dives, and sometimes breathing pure oxygen after dives to help get rid of built-up nitrogen. Just because you dive by holding your breath doesn't mean you can ignore decompression sickness, and I'll confess to once having mild symptoms that puzzled my uninformed younger self, after spending a morning repeatedly snorkelling with fins to more than twenty metres deep.

But unlike us, deep-diving mammals such as whales have evolved several adaptations to reduce the risk of decompression problems during breath-hold diving. Some species have stiff upper airways that aren't squeezed under pressure, and they can collapse the many small sac-like compartments where their lungs meet their blood supply, thereby shutting off any compressed gas from passing into their bloodstream. They also rely more on oxygen already stored by molecules in their blood and muscles, rather than using the air carried in their lungs. They may have diving behaviours that manage the risks as well, such as gradual ascents and rest periods at the surface between dives. As a result of these adaptations, there used to be some doubt about whether whales ever suffered from decompression sickness, but there is now evidence that whales can suffer from its symptoms, particularly if their normal diving behaviour is disturbed by loud underwater noises.

The autopsies of some whales that have died in some mass strandings on beaches have revealed injuries that resemble those of decompression sickness, with gas-bubble lesions in organs such as the liver and kidneys that have extensive blood supplies. Several of these strandings have been linked to naval exercises hunting for submarines with mid-frequency active sonar systems, which are different from the multibeam sonar systems that ocean-ographers use to map the ocean floor. Beaked whales in particular appear to be susceptible, and biologists are concerned that those whales may be startled by particularly loud sonar at certain frequencies, altering their normal diving behaviour and making them more susceptible to decompression sickness.

Not all strandings of whales on beaches involve decompression sickness – many have other causes, such as illness or whales becoming lost and confused by the underwater landscape around particular beaches. But there are calls for better regulations for activities that create loud underwater noises, including naval search sonars and seismic surveys that use underwater explosions to map geological structures beneath the sea floor such as oil and gas reservoirs. Current regulations typically consider the risks of direct injury to marine mammals, for example from loud sounds causing damage to their inner ears, but new regulations should also consider indirect injury from interrupting the normal diving behaviour of some whales, because of the potential for decompression sickness.

17

How do deep-sea animals find food in the dark?

Sunlight is only bright enough for algae to flourish in the upper-most layer of the ocean, down to a depth of around 200 metres in the clearest waters. Most of the food in the deep ocean starts the journey along its chain here, where some kinds of the tiny animals known as zooplankton graze on the drifting microscopic algae, and other zooplankton eat those grazers too. The poo of the zooplankton, and their dead bodies, and any other remains of the algae, become the particles known as 'marine snow' that sink into the ocean and provide food for what lives beneath. 'Marine snow' is a wonderfully poetic description for the often mushy-looking particles that drift down through the ocean. I think the term was coined by Professor Tadayoshi Sasaki of Tokyo University of Fisheries, who dived more than 9.5 kilometres deep in the bathyscaph *Archimède* in 1962. Some animals living in the interior of the ocean feed on the marine snow as it sinks, and some of them eat each other as well, but the deeper you go, the less food there is available overall from this process, because it gets eaten on the way down.

As there is less marine snow for animals to graze on at greater depths, many of the animals in the twilight and midnight zones are predators, with some amazing adaptations for catching their next meal and avoiding being eaten by others. Some of them use bioluminescence to attract their prey, such as the deep-sea angler-fish that dangle light-emitting organs on stalks that protrude from their foreheads, mimicking a small shining shrimp or another treat to tempt their prey into a toothy ambush. Other hunting animals use bioluminescence to illuminate their prey, rather than lure it. Dragonfish have organs beneath their eyes containing microbes that produce red bioluminescence. This is rare; deep-sea animals generally use blue light in biolumines- cence because it travels furthest through the water. But dragon- fish use a type of chlorophyll to enable their eyes to detect red light, which allows them to see prey illuminated by their red searchlights. Most deep-sea animals with eyes cannot detect red light, because red wavelengths are rapidly absorbed by seawater, so there is no red light remaining in the downwelling sunlight of the twilight zone. So most potential prey of a dragonfish are unaware that they have been lit up and are a target.

Bioluminescence comes in handy to avoid predators in the inte- rior of the ocean too. In the 'twilight zone' where there is still some downwelling sunlight, many animals use lights to break up their silhouettes and blend in with the light coming from above. Hatchetfish are a great example; they have two rows of organs on their bellies that produce light matching the downwelling light where they live. Hatchetfish skin also contains tiny structures that scatter light falling on them, such as predators' searchlights, which almost gives them an 'invisibility cloak', bending that light around their bodies rather than allowing it to reflect back to a predator.

As an alternative to using bioluminescence for camouflage, some animals in the twilight zone have see-through bodies to avoid casting shadows – they are like the 'invisible man' imagined by the author H.G. Wells. Several different types of animal have evolved that same solution to hide from potential predators, including some amphipod crustaceans and glass squid.

As we get down into the 'midnight zone' where there is no longer any faint downwelling sunlight, animals no longer use bioluminescence for camouflage, but blend into the darkness in other ways. Many deep-sea animals are red in colour, because red usually appears black in the midnight zone, where there are very few natural sources of red light. Red pigment may be easier for animals to produce than a true black pigment, but when it comes to camouflage it often does the same job. The blood-bellied comb jelly, for example, is a gelatinous animal with a largely transparent body, but its gut is masked with red pigment to hide any shining prey that it has swallowed. But some deep-sea fish are very black indeed, with lots of round grains of dark melanin pigment under their skin. As a result, their bodies only reflect 0.5 per cent of the light that falls on them, compared with the approximately 5 per cent reflected by most surfaces that we call 'black'.

Some animals still use bioluminescence to distract or deter predators in the midnight zone, however, and one of my favourites is the swimming green bomber worm, *Swimia bombviridis*, which lives more than 1.8 kilometres deep. Its gills have evolved into bioluminescent structures that it can eject like glowing bombs to distract predators. The worm subsequently grows new ones.

Being able to tackle large prey can help to turn any encounter with another animal into a dining opportunity. Dragonfish, for example, have an extra neck joint in their skeleton that enables them to open their jaws very wide, while viperfish have flexible jaws that can open to almost 180 degrees, and gulper eels have expandable stomachs in addition to the gaping mouths that give them their common name. Some deep-sea species appear to be active hunters, such as the Colossal and Giant Squid, but waiting for prey to swim or drift close by expends less energy than chasing it, so many deep-sea animals are 'ambush predators' instead. Tripod-fish, for example, prop themselves up on the sea floor on long pectoral and tail fins – hence their name – and sit facing upstream to detect disturbances from potential prey passing nearby.

Animals that live on the sea floor and feed by catching morsels from passing water currents face a different set of challenges.

These 'suspension feeders' need to escape a layer of very sluggish currents near the seabed, where friction slows the flow. As a result, many suspension feeders have evolved stalks to hold their filter-feeding apparatus up into better flow conditions. For example, glass sponges have rope-like stalks made of silica fibres, and there are lots of types of deep-sea corals with long stalks, such as whip corals and bamboo corals. Several other types of animal can exploit this 'scaffolding', climbing up the stalks of other species to snatch a better meal for themselves; we often find brittle stars entwined around the branches of deep-sea corals, and crustaceans perched on deep-sea sponges, for that cheeky purpose.

'Deposit-feeding' animals make their living by sucking up the particles that settle on the sea floor, extracting whatever nutrition remains in them after their journey from above. Sea cucumbers typically plough across the seabed, ingesting sediment and leaving a trail of faeces behind them. The food that falls from above is sometimes much larger than marine snow; the bodies of dead fish or larger animals from time to time sink to the sea floor. Scavenging is therefore another common feeding strategy in bottom-dwelling species, such as grenadier fish, which can detect the odour plumes from carcasses from several kilometres away and swim in efficient search patterns to home in on them.

The race to strip the flesh from any carcass at the sea floor can be intense, so any species that can make a meal from parts that others cannot use may have an advantage. Some species have therefore specialised in digesting things that are inedible to others. Bone-eating 'zombie' worms are an extreme example: they burrow into bones they find on the sea floor by secreting acid that dissolves the inorganic part of the bone, allowing the worm to consume the organic molecules that the bone contains. These worms were first seen on the bones of a whale skeleton on the ocean floor, but experiments since then have shown that they can make a living from any type of bone, as my team found out when we left half a pig carcass at 2.3 kilometres deep in the Cayman

Trough for a couple of months. When we returned to inspect the experiment all the flesh was long gone, but Dr Diva Amon found zombie worms on the pig's bones when we brought them back from the sea floor. Meanwhile, deep-sea shipworms, which are actually clams rather than worms, can digest the cellulose in any wood that reaches the ocean floor. Although wooden shipwrecks are relatively rare in general, there are areas of the world where thousands of tonnes of wood are naturally washed out into the ocean each year from rivers that run through forests, feeding the deep-sea wood-eaters.

There are some ocean-floor habitats that break the rule of food becoming less available at greater depth, where microbes grow by the process of chemosynthesis, thereby producing organic matter in the deep sea rather than relying on what has been produced by photosynthetic algae far above. Hydrothermal vents and cold seeps are the best-known examples of these chemosynthetic environments in the deep sea, where mineral-rich fluids gushing out of the sea floor provide the chemical energy that the microbes need. Several species of animals that live at vents and seeps have evolved close relationships with chemosynthetic microbes that live inside them. The large red-plumed tubeworms first seen at the Galapagos Rift hydrothermal vents initially puzzled biologists because in their adult form they have no mouths or guts; instead, they have a bag-like organ inside them, stuffed with chemosynthetic bacteria. The worm supplies these tenants with the raw materials they need for chemosynthesis: sulfide from the vent fluids, oxygen from seawater and carbon dioxide from seawater and the respiration of the animal's own cells. Getting these ingredients from the environment to the microbes inside the tubeworm's body involves some neat tricks, as sulfide is particularly toxic to most animal life. It poisons the haemoglobin molecules that usually carry oxygen in blood by attaching to them at the same spot on the molecule as oxygen, thereby preventing it from being carried. But the species of tubeworms that live at hydrothermal vents and cold seeps have evolved haemoglobin molecules with another spot that carries sulfide, without interfering with their ability to carry oxygen.

The 'scaly-foot' snail that lives at hydrothermal vents in the Indian Ocean also harbours chemosynthetic bacteria inside its body, in a sack-like gland attached to its oesophagus. The snail gets its common name from the tiny plates of an iron mineral that cover its fleshy foot, rather like armour. The species was described by my former student Dr Chong Chen during his PhD research, and he showed that it has a proportionally giant heart: if the human heart were the same, proportionally, it would be the size of one of our lungs. This huge heart helps supply the bacteria in the oesophageal gland with the oxygen and sulfide that they need, and much of the animal's anatomy appears to be modified from that of a 'normal' sea snail to support the microbes within it – the scaly-foot snail has really evolved into a carrying vessel for its bacteria.

In addition to bone-eating animals such as zombie worms, we also find animals with chemosynthetic microbes on whalebones. Hydrogen sulfide is a by-product of microbes that digest the fats inside the bones of whale skeletons on the ocean floor, and other microbes can then use that hydrogen sulfide, living in association with animals similar to those found at hydrothermal vents. Microbes that digest wood in the deep ocean also produce sulfide, similarly providing opportunities for other chemosynthetic life on 'woodfalls' as well. We are therefore beginning to realise that life finds a way to thrive in the dark using this process wherever it can.

The most common chemosynthetic microbes at hydrothermal vents use hydrogen sulfide as a source of electrons, and use oxygen to mop up those electrons at the other end of their metabolism. But some microbes at vents can use hydrogen itself as an electron source, and sulphur as the final electron-acceptor. Both of those elements may have been present at hydrothermal vents on the primordial Earth, more than four billion years ago and before any living processes had infused our atmosphere and oceans with oxygen. So a similar reaction could possibly have provided the metabolic 'spark' that got life started at hydrothermal vents on the young Earth – and could perhaps get life started anywhere else in the cosmos that has hydrothermal vents as well.

18

How do animals find a mate and set up home in the deep ocean?

Where food becomes scarce in the deep ocean, members of deep-sea species are often scattered few and far between. But to reproduce, they need to meet a member of the same species and opposite sex. Finding a mate can therefore be particularly challenging for deep-sea species, leading to some desperate measures for dating in the dark. One way to overcome the challenge is to hang on to a potential partner whenever you bump into one. So some deep-sea species exhibit 'mate-pairing' behaviour.

One of my favourites examples is the sea cucumber *Paroriza pallens*, whose cylindrical body looks a bit like a mouldy unpeeled banana lying on the sea floor. If one *Paroriza* meets another as they meander across the mud of the abyssal plain, they often then wander on side by side thereafter, staying together for when they are ready to mate. The species is actually a hermaphrodite, and when the time comes, the male parts of one individual mate with the female parts of the other and vice versa. *Paroriza* leave telltale tracks in the mud of the sea floor that record their romantic history: in sea-floor photographs, you can sometimes trace a single track winding along until it converges with another, after which a twin-track like a railway line shows the path taken by the couple in unison.

Other species take the idea of a not-so-brief-encounter to extremes. Some species of deep-sea anglerfish have 'accessory dwarf males': in these species, the male is much smaller than the female, and when boy meets girl he fuses his lips to her body in a kiss that lasts the rest of his life. The male's blood supply joins up with that of the female through his lips, and he therefore becomes dependent on her for nutrition. The female carries him around as essentially a parasitic bag of sperm, ready to fertilise her eggs when she spawns them. 'Accessory dwarf males' have evolved independently in several different types of deep-sea animals as a solution to the challenge of finding a mate: most species of bone-eating zombie worms have them, as do wood-eating deep-sea clams.

Another strategy is to be rather indiscriminate: if you encounter another animal in the deep, it can make sense to try to mate with it if it might be the same species as you. If your species has an equal sex ratio – the same number of males as females on average – then there's a fifty-fifty chance of reproductive success from having a go. Scientists have watched same-sex-mating by deep-sea octopus from the portholes of *Alvin*, and the bodies of deep-sea squid also carry evidence of such encounters. Mating in octopus and squid involves the male shooting a packet of sperm down a groove of one arm to their partner. In some species, that sperm package is like a spear, sticking to the body of their partner, hopefully close enough to female reproductive organs to fertilise eggs as they are released. Specimens of male squid fished up from the depths sometimes carry sperm packets stuck to them by other males, as testimony to their opportunistic mating strategy.

Species that reproduce by females shedding eggs and males shedding sperm into the water to mix and fertilise – a strategy known as 'broadcast spawning' – face the challenge of getting close enough together so that their eggs and sperm are likely to mix successfully. Broadcast-spawning species such as sea urchins on continental slopes appear to aggregate when it is time to reproduce, possibly triggered by seasonal changes in currents that scour the sea floor. Meanwhile at cold seeps, some clam species coordinate the release of eggs from all the females in their population, followed almost immediately by the release of sperm from all the males, using a chemical that wafts through the water to trigger that behaviour.

Once fertilised, some deep-sea animals let go of their eggs, leaving them on the sea floor or to drift in the ocean. But others, such as female Hoff crabs, carry their embryos around with them. And some species lay their eggs on the sea floor, but then keep watch over them until they hatch. Deep-sea octopuses attach their eggs to rocky surfaces and stay with them, cleaning the eggs and guarding them from predators. In May 2007, Professor Bruce Robison and his colleagues from Monterey Bay Aquarium Research Institute in California were carrying out a regular Remotely Operated Vehicle survey of the deep sea floor near their institute when they came across a female octopus that had just laid her brood of around 160 eggs on a rock face at a depth of 1.4 kilometres. They returned to the same spot eighteen times over the next four and a half years, and found the same female octopus, identified by the scars on her body, guarding her eggs as the embryos developed inside them. The mother octopus didn't appear to feed during her vigil in the dark, and her body gradually deteriorated – a testament to her maternal devotion, perhaps.

Octopus eggs hatch into offspring that resemble miniature adults in many ways, but lots of species of deep-sea invertebrates have larval stages whose bodies can be quite different from those of the adult animal, like caterpillars and butterflies. Caterpillars usually crawl around the leaves where they hatch, however, while adult butterflies flit to pastures new. In the ocean, this pattern is reversed: adult animals are sometimes restricted to one spot, particularly if they are actually glued to it, like corals, while their larval forms get carried elsewhere by currents. The ability to disperse is essential for species that live in habitat patches that do not last for ever, such as wood-devouring species colonising a sunken log that will eventually be completely consumed by them. The same is true for animals at the clusters of hydrothermal vents known as 'vent fields', because the flow of life-giving fluids does not last for ever at each field; in some areas, the life-giving fluids at one vent field can flow uninterrupted for millennia, but in other regions the flow might only last for a few decades. And the distance along the mid-ocean ridge between one vent field and its nearest neighbour can be tens or even hundreds of kilometres.

Some species, such as vent tubeworms, produce larvae that do not feed but are instead provisioned with an energy store, in the form of yolk, to sustain them on their larval journeys. As the metabolism of the larva ticks over, that internal energy store gradually gets used up, albeit perhaps slowly in the chilly waters of the deep. Careful measurements of the calorie content of a tubeworm larva's yolk supply, along with measurements of the larva's metabolism, allow scientists to estimate that the larva has around thirty-eight days to reach somewhere where it can set up home as an adult before it runs out of stored energy and dies. In the east Pacific where vent tubeworms live, the clusters of deep-sea vents are typically kilometres to tens of kilometres apart along the mid-ocean ridge, and ocean currents can carry their larvae for more than a hundred kilometres. But the current needs to be flowing in the right direction, along the ridge, otherwise the larvae can get carried away over the abyssal plain, where there are no hydrothermal vents for them to colonise.

Some other species produce larvae that can feed on organic matter raining from above during their journeys, which can allow them to survive much longer, and therefore possibly be carried much further by currents. Vent shrimps are an example of this: the bodies of their larvae look very different from those of the adults – in fact, it took biologists a while to realise they belonged to the same species. The adult vent shrimps have a light-sensing patch on their backs for an 'eye', but their larvae have two eyes on stalks like most other shrimp and bodies that are shaped to be carried more easily by currents in mid-water. Because they can feed on particles in the water, the vent shrimp larvae may be able to survive for years, drifting on ocean currents, unlike the vent tubeworm larvae with the ticking clock of their diminishing yolk supply. This difference in larval lifestyles between vent shrimps and vent tubeworms may help to explain the difference in their distribution around the world. Vent tubeworms live at vent fields that are a few tens of kilometres apart, which is where the mid-ocean ridge is rifting apart more rapidly, in the eastern Pacific. But vent shrimps are found at vent fields from the Caribbean throughout the Atlantic and into the Indian Ocean, even where the vents are hundreds of kilometres apart.

How larvae detect a suitable habitat to begin their adult lives is largely unknown, but deep ocean currents clearly carry plenty of larvae that are ready to do so. When a piece of wood sinks to the ocean floor – or is put there deliberately by scientists – it is quickly colonised, typically within a matter of weeks, by wood-devouring species such as the 'shipworm' clams. Their larvae must therefore be passing by, ready to somehow spot and then seize the opportunity to set up home. When it comes to the larvae of animals getting about in the deep ocean, there's probably a lot more going on down there than we realise.

19

What's it like to live and work at sea on expeditions exploring the oceans?

Ocean-going research ships are large pieces of scientific equipment, like radio telescopes and particle accelerators, that scientists share to carry out different investigations. In the case of research ships, separate teams of scientists use them in turn for different projects. The ships ply the oceans, coming in to the nearest convenient port between expeditions to change the science team aboard for the next voyage. One expedition might be investigating how currents swirl plankton around at the sea surface and what that means for the geochemistry of the oceans, then the next expedition might be using a Remotely Operated Vehicle to explore the geology and ecology of hot springs on the ocean floor. Each expedition can last between a couple of weeks and a couple of months depending on where it is going and what it is doing, and during that time a research ship is home and workplace to typically fifty or more people. Perhaps thirty of those aboard are the science team for that expedition, and the other twenty-plus are the ship's company who run the vessel, from the bridge officers and ship's engineers to the deck crew and the stewarding team who keep everyone fed and comfortable. The ship's company are the more permanent residents, although there are sometimes two complete crews that switch every few months, and the science team are temporary residents for their particular voyage.

Aboard a modern research vessel, there are the usual ship's spaces of the bridge and engine room, and some decks mostly devoted to accommodation. Larger ships now tend to provide individual cabins, though some older and smaller ships may have several people sharing a cabin, with a curtain around each bunk to provide personal privacy. The main deck has several laboratories that can be kitted out for the requirements of each expedition: one might be a temperature-controlled 'wet lab' where specimens of deep-sea animals can be examined at a similar temperature to that at the bottom of the ocean; another lab might be fitted with instruments to analyse the chemistry of seawater samples; and a 'dry lab' could contain computers for processing data from sonar mapping of the sea floor. There's also usually a central area, sometimes called the 'plot', where the expedition team can meet to discuss progress and plans. On the working deck outside, there are usually hangars for underwater vehicles and other equipment to deploy in the ocean, and all the cranes and winches required for that, with workshops down below to repair equipment and even build anything else if needed.

Communal areas include a 'mess' where everyone eats their meals, cooked by a team in the galley using food stored aboard for the voyage. Thanks to the creative talent of the galley team, food is usually excellent, which is so important for morale – I've always thought that the most important members of a ship's company are the captain and the cook – and fresh fruit and salad are typically available from the ship's stores for a couple of weeks after leaving port. There's also a sickbay in case of injury or illness, and for long expeditions to remote regions the expedition team might include a medical doctor, though other members of the crew are also always trained as paramedics. There's usually a lounge where people can relax when off-duty, with a library of books and films and a collection of board and card games. Ships often have some recreational deck space outside, for example high up on the superstructure away from the working deck, and some exercise equipment down below inside the hull where there is less motion of the ship, along with a laundry for everyone to use.

It's a tight-knit community, like a small village, with all the pros and cons that involves, from the camaraderie of striving together in a shared endeavour to the tensions of living and working at close quarters. Each expedition might have been years in the planning for those few weeks at sea, so the stakes to achieve the expedition's objectives are high. The 'science' part of being an ocean explorer – choosing appropriate methods to collect samples and data, and analysing and interpreting them – is only one part of the job. Success at sea also depends on all the different teams aboard ship working together effectively, and the role of the person leading an expedition – known as the 'Principal Scientist' or 'Chief Scientist' – is to be the 'glue' that helps to make that happen.

It's important for expedition team members to settle into a good daily routine aboard, to pace themselves through a long voyage. On most expeditions that I've been involved with, we've operated twelve-hour shifts, known as 'watches', for science team members. If those twelve hours are 4 a.m. to 4 p.m., then a typical day if you are working as a 'watch leader' might begin with the alarm clock going off in your cabin around 3:30 a.m. I usually switch on the reading light at the top of my bunk for a couple of minutes, bathing my face in its rays to help me come round at that unnatural hour. Then a quick jump in the shower – modern ships make their own fresh water with a reverse-osmosis system, so there's usually no shortage if we're careful – and dress. It's only a few steps via a couple of corridors and a stairway to report to the plot by 3:45 a.m., picking up a mug of coffee from an always-brewing pot in the galley on the way, in time for the handover of the watch. The fifteen minutes' over-lap between your shift and the shift going off-watch is essential for them to brief you about what's going on, what's next to do and any standing instructions from the Chief Scientist. And there are few things that become more annoying than people turning up late for the handover when you have been on duty for twelve hours.

Once the handover is complete, you can settle your watch team into the different jobs for their shift, such as routinely monitoring instruments to make sure they're working and recording their data, coordinating with the bridge officers to move the ship to a new target location, working with the deck crew to launch or recover equipment from the ocean, sorting and processing samples in the labs and analysing data gathered so far to guide further investigations.

Meals are usually taken quickly in the mess, releasing team members to eat as their work allows or if necessary asking for food to be put aside to eat later. If the expedition is using a Remotely Operated Vehicle (ROV), the twelve-hour shift might be further divided into two six-hour stints in the ROV control centre, directing where the vehicle goes and what it does at the sea floor, ensuring its video footage is carefully copied and archived and keeping good notes of what samples are being collected and where each one is stored on the vehicle ready to be transferred to the right lab teams after the dive. Operations in the ROV control centre can be very intense and more than six hours risks people losing concentration, hence the shorter time on duty there, with shift members working on other tasks in the labs during the other six hours of their watch.

After twelve hours of watchkeeping tasks, at 3:45 p.m. the other watch team should be reporting back for duty, so it's time to brief them and hand over. Then you have a few hours left in which to get some fresh air on deck if the weather allows, or get down to the gym for some exercise. You might do some laundry, and catch up with a few emails in your cabin or perhaps attempt an internet phone call to family ashore – most ships have a satellite modem shared by everyone for some internet access, though often with less bandwidth than we're used to at home. There's time for a more relaxed meal in the mess, and afterwards to wind down briefly with a book or get together with others for a movie or a board game in the lounge, before turning in to your bunk before 8 p.m. to get enough sleep to start the next day at 3:30 a.m. again.

There are no weekends or days off during an expedition: even if the ship is travelling for a couple of days between two locations for our work, we still keep watches and record data from instruments all the time, for example mapping the ocean floor with the ship's multibeam sonar system. So a good daily routine is important, because some days will be stretched by dealing with unexpected discoveries or problems that could otherwise strain your stamina.

The role of Chief Scientist is different, although trying to establish a daily routine at sea is still important in that role for the same reasons. The Chief Scientist's job starts a couple of years before the expedition sets out, working with a logistics team for the ship to plan what equipment needs to be aboard and how to get it there. The UK's research ships usually return to the UK once per year for annual refit, and that offers a chance to load some equipment aboard for forthcoming expeditions, if there is space in the ship's hold to store it. But much of the equipment has to be freighted to the port from where the expedition will begin.

An expedition heading out into the Southern Ocean from a port at the tip of South America, for example, might need to send equipment there via a trans-shipping hub in the Caribbean, depending on what 'slots' are available for containers on different shipping routes. And a deep-sea ROV system might need five shipping containers for the vehicle, its launch system, its spares and its control centre. So freighting equipment to the starting port for an expedition, loading the containers on and off the ship in the port, and sending them back from the final port afterwards can be a major cost, sometimes exceeding US$250k, but that's still cheaper than bringing the research ship halfway across the world to its home country in-between every expedition.

If an expedition is going to be working inside the Exclusive Economic Zone (EEZ) of another country, which is usually a 200-nautical-mile zone beyond that country's coastline, then the Chief Scientist also has to prepare a request for diplomatic clearance to send to the government of that country. The request needs to explain the research that the expedition would like to carry out and why it matters, and indicate how scientists from that country will be involved. If the research is deemed sensitive to the economic interests or security of the country, the request might be refused, in which case it's back to the drawing board for the Chief Scientist. A ship straying into another country's EEZ to gather data without diplomatic clearance would create an international incident, and approval can take several months to be granted, so those formal requests need to be prepared and lodged at least a year in advance.

In preparation as a Chief Scientist, you also need to consider who will make up the expedition team. To achieve the goals of the expedition, you may need several different types of scientists aboard, for example bringing together geophysicists, geologists, oceanographers, geochemists and marine biologists, along with specialist engineers to operate underwater vehicles or other equipment. To work around the clock, the expedition team needs to be divided into different shifts, but do you have enough of each type of specialist in each shift for the tasks that you anticipate? Is there a good mix of experienced hands and next generation in each of your teams, and how well do you think the different personalities of individual team members will work together under pressure?

Finally, you might draft an initial plan for the expedition: what locations the ship will visit, with a rough timetable of activities such as how many underwater vehicle dives you anticipate at each location, some thoughts about the configurations that the vehicle will need to collect different samples and data and what other tasks the ship will need to complete to achieve the expedition goals. That plan will need to be discussed with the other scientists involved, to show how it addresses their individual ambitions and expectations for the expedition. And your 'Plan A' will certainly change – in fact, you can expect to get through many letters of the alphabet in alternative plans during the expedition, as new constraints become apparent, problems arise and conditions change. Being able to identify the opportunities of a better plan and pivot to it rapidly is an important skill, but it's good to have a starting point, and by the time you walk up the gangway to board the ship as Chief Scientist, you have already invested a great deal of time and thought in the expedition.

Once at sea, the role of the Chief Scientist involves communication and decision-making: ensuring everyone in your expedition team knows what they're doing at every moment, and getting the ship's company involved with and motivated about the expedition's goals. There's usually a daily meeting for the Chief Scientist with the ship's captain and senior staff, to review how things are going, present and refine plans for the next twenty-four hours and consider anything that may be developing beyond that, such as changes in the weather forecast. You also need to listen to the different teams aboard ship about how their work is going, and weigh up the needs of individual scientists or groups with the overall goals for the expedition. A routine daily meeting of the expedition team is therefore essential, ideally timed near a handover between watches so that both shifts can attend. All of that feeds into developing detailed plans for the next few days ahead, including contingencies that consider possible changes in conditions and making sure those plans are clearly and continually communicated to everyone. Meanwhile, it's also good to try to join in with some watchkeeping where possible, to keep a feel for what's going on and how people are getting on. If you're occasionally able to make cups of tea for the watchkeeping shift, it's usually a sign that you're on top of things.

The Chief Scientist deals with strategy and the uncertainties of the future, so that watch leaders have a clear set of instructions and options for everything to run smoothly. Consequently, there can be less of a predictable routine when you're occupying 'the big chair' for an expedition, but of course you still need sleep to function at your best. When something happens while you are asleep, for example a weather front moving in or a piece of equipment breaking down, hopefully you have anticipated it and discussed the options with your watch leaders, and they can often take appropriate action without needing to rouse you. But sometimes they may face an unexpected choice with potentially far-reaching consequences, which they need to pass to you. So the phone in your cabin can wake you at any time, and there's no harm in that when you trust the judgement of your watch leaders.

As Chief Scientist I like to define four fundamental 'states' for the ship, with standing instructions that the current 'ship's state' should be clearly displayed for all to see, and the times of changes between those states recorded in the watchkeeping log. The four states are: 'equipment deployed', when we have a vehicle such as an ROV working in the water, or an instrument running, gathering data or samples; 'deck work', when the deck crew or expedition engineers are busy preparing or fixing a piece of equipment and nothing else can be deployed in the meantime; 'in transit', when the ship is travelling to a new location and therefore can't deploy equipment, which includes poor weather when all we can do is ride out the wind and waves; and 'idle', when the ship is not deploying equipment, preparing equipment or travelling – an indication that someone somewhere doesn't know what to do next. That someone could be you as Chief Scientist, if you are deciding what to do in response to something unexpected that has just happened. But it could also be a watch leader, or the officer on the bridge, if there is a breakdown in communication. The ship should instead move seamlessly between tasks in the other three states, without passing through any 'idle' periods.

At the end of an expedition, the watchkeeping log reveals how much time the ship has spent in each of the four states, and the overall goal during the expedition is to avoid any time logged as 'idle', and maximise time spent with 'equipment deployed'. Your expedition is years in the making, and hundreds of thousands in its funding, with one chance to achieve its goals, so every moment counts and having an 'idle' ship is inexcusable. There's also an optimum use of the ship to try to achieve in your plans: for example, it may be possible to carry out 'deck work' preparing a piece of equipment while the ship is also 'in transit' to a new location. And your plans involve people, not just machines: for example, if the ROV has just collected a lot of geological samples on a dive, then the geology team will need some time to process those samples before you swamp them with any more, so switching to another activity may be a good idea. Furthermore, some activities may be more restricted than others by weather: sonar mapping can continue in rough seas, for example, whereas launching the ROV cannot. All these kinds of constraints combine to make your planning something like a sliding-block puzzle, and it's very satisfying to come up with an efficient solution.

Ultimately the Chief Scientist is responsible for the expedition achieving its goals, justifying the investment of funding for it. A month-long expedition using deep-sea vehicles might cost upwards of US$1 million, with the cost of freighting equipment to and from the ship, port fees, the ship's fuel bill and provisioning, the salaries of all those aboard, the travel of expedition team members to and from the ship and the costs of specialist supplies and equipment that the expedition will use. Most expeditions are paid for through grants for research projects that scientists propose and compete for, usually funded by governments, or sometimes private foundations interested in exploring the oceans. But wherever the funding comes from, the Chief Scientist's name is on the line for delivering the results from that investment.

20

*What have been your favourite expeditions
so far, and which historic voyages of ocean
exploration do you most admire?*

Each expedition is unique, and working aboard the research
ships of other nations offers a chance to explore different
cultures as well as the oceans. At the start of my career I enjoyed
getting to know and working with a Russian ship's crew just
after the end of Soviet rule, and then for my next expedition
enjoyed a very different experience aboard a ship with a US
Navy unit. But if I had to pick favourite expeditions in terms of
discoveries, it would be a pair of expeditions in the Southern
Ocean in 2009 and 2010, which were part of a UK project to
explore hydrothermal vents near Antarctica. The whole five-
year project was led by my long-time mentor Professor Paul
Tyler at the University of Southampton, who has spent more
time than any other British biologist in history diving in Human-
Occupied Vehicles to study deep-sea life first-hand. Paul also led
the proposal that equipped the UK with its first deep-diving
Remotely Operated Vehicle (ROV) for science, and I am one of
many in the 'family' of deep-sea scientists that he raised who
owe our careers to him.

The first of the two expeditions was aboard the RRS *James
Clark Ross*, led by Dr Rob Larter of the British Antarctic Survey. It
was a reconnaissance mission for the overall project, investigating
where other scientists had previously found hints and clues of
possible colonies of chemosynthetic life in the Southern Ocean.
Some of those hints and clues proved correct, others did not and
there were some other unexpected discoveries along the way, so
the expedition was the usual rollercoaster with highs of exhilara-
tion and lows of frustration, but it felt like a voyage of

exploration in its purest form, seeing parts of our planet for the first time with a camera system that we towed beneath the ship.

The second expedition returned to investigate the astonishing colonies of creatures revealed by the first expedition, using the UK's ROV aboard the RRS *James Cook* and led by Professor Alex Rogers of Oxford University. All the animals that we found thriving around the hydrothermal vents in the Antarctic were new to science – more than thirty previously unknown species. And although I had worked with several other ROVs before, we really put this one through its paces, spending more than 300 hours working at the sea floor down to 2.6 kilometres deep in the challenging conditions of the Southern Ocean, which showed what modern technology can achieve.

Modern expeditions are very different from the early voyages of ocean exploration, as we no longer have to contend with the hardships faced by mariners in the past. During the round-the-world expedition of HMS *Challenger* from 1872 to 1876, seven people died, twenty-six were left behind in port hospitals around the world and several of the crew deserted. But the voyage of HMS *Challenger* is one of the expeditions from history that I admire the most, for the scale of its ambition and because it established the multidisciplinary approach to exploring the ocean that we still follow today. Each of the 362 locations where the ship stopped to make measurements and collect samples were roughly equally spaced along its route to provide global coverage, and the same methods were attempted at each location to collect combined data about physics, chemistry and biology. As a result, the 'sealed volume' of the ocean's interior, as Matthew Fontaine Maury had described it just two decades earlier, was opened up to reveal its secrets like never before.

Plans for the *Challenger* expedition built on foundations of knowledge and experience from other voyages, such as scientific expeditions in the north-east Atlantic by HMS *Lightning* in 1868 and HMS *Porcupine* in 1869 and 1870. Both those naval vessels were made available for science at the request of the Royal Society, and their expeditions dredged up specimens of marine life from as deep as 4.4 kilometres, south-west of Ireland. They also investigated the chemistry and physics of the deep ocean, for example revealing layers of deep water with different temperatures in the channel between Scotland and the Faroe Islands. The successful combination of different branches of sciences to understand how the oceans work prompted those involved to hatch even bigger plans for a round-the-world voyage, which would study all aspects of the depths of every ocean.

On 8 June 1871, the scientific journal *Nature* reported a talk about ocean exploration by William Carpenter at the Royal Institution in London, and the article included a rallying call for further British endeavours:

> . . . *having shown other nations the way to the treasures of knowledge which lie hid in the recesses of the ocean, we are falling from the van into the rear, and leaving our rivals to gather everything up. Is this creditable to the Power which claims to be mistress of the seas?*[18]

Having thumped that tub of national pride, Carpenter mobilised leading scientists in November 1871 to lobby the British government for what would become the *Challenger* expedition. The names of those involved read like a who's who of nineteenth-century science in Britain, with Charles Lyell speaking for geology, Thomas Henry Huxley for biology and John Tyndall for physics, with further support from polymath William Thomson. Persuaded by promised vistas of scientific discovery, and with an eye on the strategic importance of understanding more about the ocean floor for laying submarine telegraph cables, the British government backed the project, which cost £200,000 (US$265,000) at the time, equivalent to about £22.4 million (US$29.5 million) today.

At the age of fifty-nine, Carpenter was too old to take part in the voyage, which was led by 41-year-old Charles Wyville Thomson, a fellow veteran of the scientific expeditions aboard HMS *Lightning* and HMS *Porcupine*. The Admiralty assigned HMS *Challenger*, a 60-metre corvette with auxiliary steam power in addition to her sails, for the expedition. All but two of her guns were removed so that she could be fitted with tried-and-tested technology for deep-sea sounding and dredging, including nearly 200 kilometres of assorted ropes and lines, on-board laboratories for curating samples and a new-fangled photographic darkroom. The aft cabin of Captain George Nares was divided in two to accommodate the Principal Scientist with equal prominence. Five scientists and one artist joined the ship's company of twenty-one officers and 216 crew in Sheerness, and the ship finally left British shores from Portsmouth on 21 December 1872.

The *Challenger* expedition returned nearly three and a half years later, arriving in Spithead on 24 May 1876 having travelled across more than 127,000 kilometres of ocean, which is nearly one-third of the distance from the Earth to the Moon. One hundred years later, the lunar module of *Apollo 17* was aptly named *Challenger* in honour of that voyage.

More than 100 scientists ashore analysed the data and samples collected by HMS *Challenger*, and the results of their work filled fifty volumes published between 1877 and 1895. John Murray, who was a naturalist aboard the expedition, oversaw the complete publication of these '*Challenger* Reports' following Charles Wyville Thomson's death in 1882, and the reports include the descriptions of more than four thousand new species collected during the voyage. As Wyville Thomson wrote of the voyage afterwards: 'Strange and beautiful things were brought to us from time to time, which seemed to give us a glimpse of some unfamiliar world',[19] and the *Challenger* expedition helped to spur other oceanographic expeditions in the late nineteenth century, such as those by the *Blake* and the fifth USS *Enterprise*

(United States), the *Travailleur* and the *Talisman* (France), the *Ingolf* (Denmark), the *Hirondelle* and *Princess Alice* (Monaco), the *Valdivia* (Germany) and the *Siboga* (Holland). As we approach the 150th anniversary of the voyage, we have to wonder what, with the tools that are now available to us, a modern-day equivalent of the *Challenger* expedition could achieve for our understanding of the ocean.

21

How do our everyday lives affect the deep ocean?

Charles Saxon, whose stylish drawings illustrated the *New Yorker* magazine for several decades, once produced a cartoon depicting a matronly group of Manhattanites chatting at a coffee morning, with the caption 'I don't know why I don't care about the bottom of the ocean, but I don't.'[20] That sentiment is quite understandable: we care most about things that are connected to us in some way, and the ocean depths seem a remote and alien place. But our lives are far more connected to the deep ocean than we might realise.

Most of us now use the deep ocean every day, but probably without knowing it. If you have been online today, sharing any information with computers on other continents by visiting websites or sending messages via social media and old-fashioned email, then you have used the deep ocean. The information that you have sent and received has travelled across a modern network of fibre-optic cables on the ocean floor, which now carry more than 99 per cent of telecommunications traffic around the world. Similarly, when we phone someone on another continent, our conversation no longer bounces off satellites orbiting the Earth – it is now also carried by those fibre-optic cables across the ocean floor, which provide far more capacity than satellite links. And if you are reading a digital version of this book, it seems somehow fitting to think that it might have travelled across the ocean floor to you when you downloaded it.

We have actually come full circle in this use of the deep ocean. Sea-floor telegraph cables revolutionised communication in the late nineteenth century, enabling political and commercial conversations to take place between continents in mere hours, rather than weeks via messages on ships. When the first transatlantic cable was laid in 1858, *The Times* newspaper optimistically declared that 'The Atlantic is dried up, and we become in reality as well as in wish one country',[21] and Kipling's poem 'The Deep-Sea Cables' from 1893 ends with these lines:

> *Joining hands in the gloom, a league from the last of the sun.*
> *Hush! Men talk to-day o'er the waste of the ultimate slime,*
> *And a new Word runs between: whispering, 'Let us be one!'*[22]

But while the first transatlantic telegraph cable could carry between ten and twelve words per minute, stuttered along it in Morse code, the latest transatlantic fibre-optic cable can stream 72 million high-definition movies simultaneously. That's around 16 trillion times more information, in just 160 years.

Today's communication cables are a relatively passive use of the deep ocean, however, not exploiting its resources and – fortunately – having little effect on deep-sea environments. There are much greater human impacts on the deep ocean, of course.

Fishing of deep-sea species grew in the late twentieth century, as more of the fish stocks in shallow water became fully exploited or over-exploited. But some deep-sea fish are particularly vulnerable to overfishing, as they can be long-lived, slow-growing and take years to reach reproductive maturity. One example is the orange roughy, *Hoplostethus atlanticus*, which can live for more than a century and does not start to reproduce until it is in its twenties. Adults of the species congregate at seamounts and submarine canyons, where they can therefore be caught relatively easily. Fishing of orange roughy began in the 1970s, and by the late 1980s the annual global catch was over 90 thousand tonnes; then stocks and catches declined rapidly. But in the waters around New Zealand, where careful management plans have been devised and put in place, several stocks have since stabilised.

It's not just the target species of deep-sea fisheries that can be affected by fishing, however. Bottom-trawling can damage fragile deep-sea habitats, with a corresponding impact on many of the other species that live there. For example, the slow-growing stony corals on continental slopes and seamounts provide a habitat for many species of deep-sea fish and invertebrates. But if the fish are targeted by bottom-trawling, the coral is also destroyed, and will take centuries to regrow. In 2016 the European Union therefore agreed to ban bottom-trawling at depths greater than 800 metres in European waters, and to close areas deeper than 400 metres where there are vulnerable habitats such as coral and sponge beds.

As our use of living resources in deeper waters has grown in recent decades, so has our use of non-living ones such as oil and gas, which can now be extracted from prospects in much deeper water than before. In 2016, the *Maersk Venturer* drillship set a new record for the deepest sea-floor depth of an offshore oil well, at 3.4 kilometres deep off the coast of Uruguay. Instead of rigs fixed in place by legs to the sea floor, deep-water operations use drill-ships or semi-submersible platforms, which are stabilised by pontoons below the sea surface. The ship or platform is kept in position above a wellhead on the sea floor, often using 'dynamic positioning', where computer-controlled thrusters respond automatically to data from global positioning system satellites. In one sense, subsea oil wells can help marine conservation, because they result in small 'marine protected areas' around them, where fishing is prohibited to avoid snagging the wellhead on the sea floor. But when accidents occur during oil and gas extraction in deep water, the results can be catastrophic for people and marine life.

Deepwater Horizon was a semi-submersible platform undertaking the drilling of a new well in the Gulf of Mexico. On 20 April 2010, it was completing its work when it suffered a 'blowout'; a slurry of drilling mud, methane gas and water erupted up onto it. The gas exploded, killing eleven people aboard and eventually sinking the platform. Meanwhile, oil spilled from the wellhead at a depth of 1.5 kilometres on the sea floor, until it was eventually capped on 15 July, eighty-six days later. The amount of oil spilled is estimated at 780 million litres – the largest oil spill in the history of the petroleum industry. Its impact permeated the ecosystem, affecting species from deep-sea corals near the wellhead to plankton, tuna and dolphins. The operator agreed to pay more than US$18 billion in fines and penalties, which was the largest corporate settlement in US history.

As well as our pollution, our rubbish also ends up in the deep ocean. In the logsheets that we use to record observations at the sea floor during expeditions, there are several categories for noting any human impacts that we encounter. On an expedition a few years ago, to pass the time on watch while our ROV was descending to the ocean floor one of my research students asked me which of the categories I had seen before in the deep sea. The answer was all of them. Discarded fishing nets? Yes, on a remote seamount in the Indian Ocean. Discarded long-lines from fishing? Yes, at two kilometres deep in the Southern Ocean. Plastic and other rubbish? Yes, a bin liner at 2.3 kilometres deep near hydrothermal vents in the Caribbean, along with a beer can next to it. Scrap metal? Yes, a tangle of discarded pipes at around three kilometres deep on the Mid-Atlantic Ridge north of the Azores.

Human-generated rubbish unfortunately has a long history in the deep ocean. In the age of steamships, vessels dumped the remains of burned coal, known as clinker, from their engine rooms. Clinker changed the nature of the sea floor along some well-travelled routes, transforming the seabed from soft mud suited to burrowing forms of marine life into cobbled areas favouring other life forms that anchor to hard surfaces like that provided by the clinker.

But at the time when our great-great-grandparents were shovelling clinker over the sides of ships, they only had hazy notions about the depth of the oceans and what was going on down there. Plastic has since replaced clinker as a common contaminant of the deep ocean, however, and ignorance can no longer be an excuse. In addition to the plastic bags that I have seen around the ocean floor, smaller pieces of plastic known as microplastics have become ubiquitous in the ocean. Dr Lucy Woodall of Oxford University and her colleagues have shown that sediment samples from the deep seabed contain 10,000 times the concentration of microplastic particles that are floating in surface waters, because the particles drifting at the surface eventually sink and accumulate at the ocean floor. And Dr Alan Jamieson of Newcastle University and his team have found that the guts of animals from the bottom of deep ocean trenches contain microplastics, and they have organic pollutants in their tissues.

Even if microplastics are not toxic to deep-sea life, they could still have an impact on it; at the moment, we don't yet understand exactly what the impacts are likely to be. Imagine a deposit-feeding animal such as a sea cucumber, ploughing across the surface of an abyssal plain and ingesting all the particles in its path. If a proportion of the particles that it swallows are now microplastics, with no nutritional value for the animal, then the food that it actually needs will be diluted by those worthless particles. Although the sea cucumber is unlikely to starve as a result, it may end up with less energy to use in reproduction, or to expend in evading predators by swimming up off the seabed. Either of those outcomes could alter the fate of its population, or the balance of its interactions with other species.

Even smaller than microplastics, the carbon dioxide molecules from burning of fossil fuels also have an impact on the deep ocean. As the carbon dioxide that we pump into the atmosphere dissolves into the ocean, it makes the seawater less alkaline – a problem known as 'ocean acidification'. Where the waters of the deep ocean are relatively 'young' – only just starting out on the centuries-long journey of ocean circulation from the surface into the deep – the recent change in the chemistry of water sinking from the surface has made the Carbonate Compensation Depth (which we met in the Question 11 chapter) shallower than pre-industrial times in some areas, such as the Southern Ocean and the north Atlantic. That in turn affects the marine snow sinking into the deep, as carbonate skeletons of deceased lifeforms in those particles will dissolve more quickly. Less alkaline water may have other direct effects on living creatures in the deep, but they're not yet understood.

Rising global temperatures reduce the amount of oxygen that can dissolve in seawater too, and climate change also weakens ocean circulation by altering sea ice formation and wind patterns at the poles, with the end result that low-oxygen zones may be expanding in the deep ocean, which will lead to some redistributions of marine life around them. As we seem to be waking up to the issue of plastic waste flowing into the ocean, perhaps we can apply the same attention to the flow of carbon dioxide molecules into our atmosphere, which also affects the ocean. Scientists have been warning governments about that issue for decades, of course, and there have been international agreements to tackle the problem, but time is running out to make changes that are necessary to reduce major impacts.

Our modern lives have also created a huge demand for metals, which could lead to impacts from deep-sea mining in the future. For example, a conventional family car contains perhaps around thirty kilograms of copper, but a 'hybrid' car that produces lower greenhouse gas emissions may need around eighty-five kilograms. Meanwhile, consumer technology depends on Rare Earth Elements, such as indium, used in touch-screens, tellurium, needed for fibre-optics and improved solar cells, and neodymium for magnets in wind-turbines. Deposits of these rare earth elements on land are largely controlled by single nations, so the hunt is on for sources elsewhere to avoid potential monopoly, as well as to meet growing demand. The mineral deposits of the ocean floor could fit the bill, and so deep-sea mining has become a nascent industry.

There are currently three main types of mineral resources being targeted for deep-sea mining. Polymetallic sulfides, rich in copper and gold – in some cases with ten times the copper concentration of ores mined on land – are found in the mineral spires that form at hydrothermal vents. Some areas of the sea floor out on the abyssal plains are littered with small potato-sized deposits called manganese nodules, which are a potential source of Rare Earth Elements. And the summits of some seamounts are ringed with 'cobalt crust' deposits, which can also be rich in Rare Earth Elements, as well as cobalt, which is currently in demand for electric car batteries.

Each of those deep-sea environments is very different in size and the kind of marine life that lives there. The total area of all the active hydrothermal vents currently known in the deep ocean is about 50 km^2, which is about half the size of Disney World in Florida. Hydrothermal vents are only inhabited by a few hundred animal species that can cope with their conditions, but most vent species are not found in any other types of habitat, and each island-like set of vents is only inhabited by a few species from that total. In contrast, the abyssal plain area of the eastern Pacific where manganese nodules are found spans more than 6 million km^2 and contains thousands of species, but typically in sparse populations compared with marine life in the other habitats. And there are around 9,000 large seamounts worldwide, each with an average area of 700 km^2, and they are home to species such as very slow-growing deep-sea corals.

Mining each type of deep-sea habitat will involve different techniques: the polymetallic sulfides of hydrothermal vents will be strip-mined; fragmentation mining will be required for cobalt crusts on seamounts; and manganese nodules will be sucked up from the sea floor. Each type of habitat is likely to respond very differently to these different impacts of deep-sea mining. In January 2018 the European Parliament called for an international moratorium on deep-sea mining until there has been sufficient further research to understand what effects it is likely to have on marine environments and their biodiversity, but the United Nations organisation that regulates deep-sea mining in international waters continues to press ahead with developing it.

22

What can we learn from the species that live in the deep ocean?

Space exploration has produced spin-off lessons and innovations for our everyday lives, from cordless vacuum cleaners to freeze-dried food, but what has exploring life in the deep ocean ever given us?

Here are a few examples of spin-offs from studying the species that live in the deep ocean, some of which are already in use and others that have potential for the future. Let's start with one that we may have encountered without realising it. The ocean pout, *Zoarces americanus*, is a species of fish that lives on the continental slope of the Atlantic coast of North America from Delaware up to Labrador, where waters are often cold. The fish produces anti-freeze proteins that curb ice crystals from growing in its blood and damaging its cells. Ice crystals also cause problems for ice-cream manufacturers: if ice cream partially thaws and refreezes, for example during shipping, its ice crystals inevitably get larger, which makes its texture less appealing. And if ice crystals can also be kept small during manufacture, it may be possible to make smooth ice cream using less fat.

Scientists managed to identify the gene that the ocean pout fish uses to make its antifreeze protein, and transfer that gene into yeast cells, which can then be grown in vats to produce the protein in industrial quantities. Extracting the antifreeze protein and removing any remnants of yeast cells from it has provided ice cream manufacturers with a new ingredient, called ice structuring protein, to tackle their problems with ice crystals. Ice structuring protein is identical in form to the protein occurring naturally in the ocean pout's body, and only a tiny amount is needed for its beneficial effect. But there is no fish, or any yeast, in the ice cream that contains the ingredient. It has been approved by food regulators in most regions of the world, including the US, Australia and New Zealand, and the European Union. So next time you are enjoying a particularly creamy frozen treat, have a look to see if ice structuring protein or 'ISP' features on the ingredients list.

Elsewhere in the deep sea, glass sponges grow beautifully intricate cage-like skeletons from silica fibres, giving rise to some evocative species names such as the Venus Flower Basket. At their base, the sponges grow special glass fibres to anchor themselves to the seabed, and those anchoring fibres have a layered structure like those of the optical fibres used in telecommunications cables. The sponge's glass fibres have natural fibre-optic properties, similar to those of our manufactured optical fibres, but they are tougher; in particular, their combination of inorganic and organic materials is very effective at stopping cracks from growing, a property not shared by man-made optical fibres, which fracture more easily. Unlike our optical fibres, which are made at high temperatures, the sponge grows its fibres in the low-temperature conditions where it lives, which may possibly contribute to making them tougher. So studying how the sponge makes its glass fibres may help us to develop better fibre-optical materials.

The scaly-foot snail also offers insights for materials scientists, specifically from the structure of its shell. The innermost layer of the snail's shell is made of calcium carbonate, which is typical for a sea snail. On top of its inner shell, there is an organic layer, similar to the organic coating that usually covers the outside of snail shells. But the scaly-foot snail's shell has an additional outer layer, containing crystals of an iron sulfide mineral called greigite embedded in softer organic matter in an organic matrix.

Together, these different layers provide a unique combination of fracture resistance and stiffness. If something strikes the outer layer, the iron sulfide crystals scattered over its surface can shear away to absorb some of the force, unless the blow is exactly perpendicular to the curved shell at its point of impact. But any cracks created in the outer layer also branch out quickly in-between the iron sulfide crystals, and a crack consisting of many fine branches is less damaging to the shell. The compliant middle layer of organic matter in the shell absorbs some of the impact as well, and helps to stop any cracks from growing in the innermost calcium carbonate layer, which provides the shell's overall structural stiffness. So in terms of its overall performance in resisting damage, the whole of the scaly-foot snail's shell is greater than the sum of its parts, thanks to the interactions between its components.

Having revealed the unique properties of this deep-sea snail shell through experiments, materials scientists have been working to incorporate its principles into better crash helmets, body armour, pipeline protection and even scratch-resistant paint.

The myriad of microbes that thrive in different ways in the deep ocean have great medical potential, and one of them has already given rise to a potential new treatment for prostate cancer. Some deep-sea bacteria contain a type of chlorophyll known as bacterio-chlorophyll, which they use to respond to faint light sources in the deep ocean. One type of bacteriochlorophyll from the deep sea has been developed into a compound called padeliporfin, which produces damaging forms of oxygen inside cells when it is exposed to light. Padeliporfin can therefore be injected into the bloodstream of patients with prostate cancer so that it travels to the tumour, and then a flash of laser light sets off its reaction, which kills the cancer cells.

The results of trials published in *The Lancet Oncology* in 2017 showed that 49 per cent of patients treated with 'vascular-targeted photodynamic therapy' using padeliporfin went into complete remission, and the treatment is less invasive and has fewer lasting side effects than surgery or radiotherapy for localised prostate cancer. It has taken decades to develop and modify the original deep-sea bacteriochlorophyll molecule to make it safe and effective as a light-sensitive drug, and its use for treating prostate cancer is currently being reviewed by the European Medicines Agency, so it will be a few more years before it may become widely available. But researchers are investigating whether it may also be useful for treating other solid tumours such as liver and breast cancer.

Ocean microbes, and animals such as sponges that have to contend with microbes living on or inside them, are also being investigated for possible new compounds and approaches to tackle the growing problem of antibiotic-resistant diseases. Antibiotic resistance is one of the most pressing threats to our population, with the annual total of deaths from antibiotic-resistant diseases predicted to exceed those from all cancers by 2050. We desperately need to stay one step ahead of infections that are evolving to resist our current antibiotics, and come up with new ways of disrupting them that are harder to resist.

The biodiversity of the oceans is a library of the ingenuity of nature, and we can learn a lot from exploring its shelves. But it feels like we're also playing with matches in that library, through our other actions in the oceans, and we risk losing some of the books of life before we have had the chance to consult them.

23
Who owns the deep ocean?

The prospects of mineral wealth beneath the waves have already redrawn the geopolitical map of the world. When it comes to claiming undersea territory, the rules are defined by the United Nations Convention on the Law of the Sea (UNCLOS), which came into force in 1994 and has now been ratified by 167 separate nations and the European Union. Under the terms of UNCLOS, nations automatically get 200 nautical miles of sea floor around their coastline as an Exclusive Economic Zone, with rights over non-living seabed resources such as oil and gas and other minerals. And if that 200 nautical miles collides with that of another nation, then the sea-floor territory gets divided down the middle between them.

Tiny island specks from a nation's history therefore become the centres of rather larger areas of seabed territory. The UK Overseas Territory of Pitcairn is a good example: a remote Pacific island, originally settled by mutineers from HMS *Bounty* and with fewer than 100 inhabitants today, it is now the centre of British sea-floor territory covering 834,000 km^2 around it. Add in other UK Overseas Territories such as Ascension, the Falklands, South Georgia and the South Sandwich Islands and the Chagos Archipelago, and the UK's overall sea-floor territory becomes twenty-seven times larger than all of the nation's land area, covering 6.8 million km^2 of seabed in total. When I once pointed this out to a UK science minister, to highlight the importance of maintaining a deep-ocean capability, they quipped 'Yes, but there aren't any voters down there!' – a rather dismal reply from a politician even in jest, as there are undoubtedly things down there that voters care about, such as economic opportunities and environmental responsibilities.

Seafloor Territories

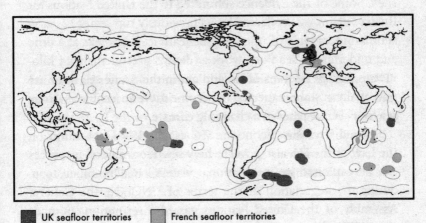

| UK seafloor territories | French seafloor territories |

France is the biggest 'winner' in sea-floor territory under UNCLOS, with the world's largest collective EEZ around its territories, covering more than 11,691 million km². Meanwhile, the ownership that EEZs confer for oil and gas reserves, and other sea-floor minerals, is also the reason for tensions between nations over islands, particularly in the seas around South East Asia. And under the rules of UNCLOS, it is also possible for nations to extend their EEZs beyond 200 nautical miles, if they can show that the seabed there has a geological connection to their continental shelf. Such claims involve submitting a detailed case to the United Nations with geological evidence from seabed rock samples and mapping of sea-floor terrain, which several nations have done so far: the UK, for example, has submitted a claim for more territory around Ascension in the Atlantic. But perhaps the most notable claim is that of Russia in the Arctic

Ocean, where Russia argues that the Lomonosov Ridge, which cuts across the ocean basin almost to the North Pole, may be considered a geological extension of the Siberian continental shelf. Some of the evidence submitted to the United Nations for Russia's claim was collected by its previously top-secret *Losharik* submarine, capable of operating beneath ice for weeks at a time and undertaking sea-floor work at depths greater than 2.5 kilometres. Likely reserves of oil and gas in the Arctic sea floor are potentially at stake, expected to become more accessible as Arctic sea ice cover declines in a changing climate.

Beyond Exclusive Economic Zones, the UN Convention on the Law of the Sea also governs how sea-floor mineral resources can be exploited in 'international waters'. In December 1970, during the negotiation of the terms of UNCLOS, the General Assembly of the United Nations passed a resolution (number 2749/XXV) that:

> the area of the seabed and ocean floor and the subsoil thereof, beyond the limits of national jurisdiction, as well as its resources, are the common heritage of mankind, the exploration and exploitation of which shall be carried out for the benefit of mankind as a whole.

The principle of 'common heritage of humanity' for sea-floor mineral resources in the 'High Seas' area beyond national jurisdiction began in 1967, when Dr Arvid Pardo made a radical proposal as Malta's delegate to the United Nations. At that time, research estimated that the ocean floor held almost limitless mineral resources, with enough of various metals down there to meet global demand for many millennia. Pardo saw those potentially vast resources as an opportunity to redress some of the injustices of history, and he proposed that benefits from exploiting that mineral wealth should be shared, particularly among developing countries. His idea was a noble one: as our 'common heritage', the enormous resources of the deep ocean could help to lift those in poverty to a better future, and thereby create a more equitable global society of nations. Pardo's principle was therefore adopted in the goals of UNCLOS, which are to 'promote the peaceful uses of the seas and oceans, the equitable and efficient utilization of their resources, the conservation of their living resources, and the study, protection and preservation of the marine environment' and thereby contribute to 'the realization of a just and equitable international economic order which takes into account the interests and needs of mankind as a whole and, in particular, the special interests and needs of developing countries' (UNCLOS Preamble, 1982).

Unfortunately, we now know that the assumption about the deep sea being able to meet global demands for various metals for thousands of years is mistaken. For example, all the world's known active and inactive deep-sea hydrothermal vents contain a total of around 30 million tonnes of copper and zinc, while around 19 million tonnes of copper and 12 million tonnes of zinc are mined each year on land, according to a recent assessment led by Dr Cindy Van Dover of Duke University. Rather than a seemingly bottomless treasure chest that could be shared to create a more equitable world, some of the mineral deposits of the ocean floor are a more limited resource than Pardo envisioned.

Nevertheless, the International Seabed Authority (ISA) of the United Nations was created in 1994 to administer future mining of the ocean floor in international waters, and determine how the profits should be dispersed to achieve the 'common heritage' goals enshrined in UNCLOS. With a permanent headquarters in Kingston, Jamaica, the ISA has a Legal and Technical Committee of around two dozen people – lawyers, economic geologists and a few environmental scientists – who are currently responsible for overseeing the awards of licences to explore for sea-floor minerals in international waters, and considering the environmental impacts of those activities. Right now the ISA is also drafting the rules that will govern the future industrial extraction of sea-floor minerals in the High Seas beyond national jurisdiction, and that area of deep ocean covers around 45 per cent of our planet's surface.

There are certainly challenges ahead for the International Seabed Authority, particularly in how it balances its objective of delivering financial benefits from deep-sea mining with responsibilities for environmental stewardship. At the moment, for a mineral exploration licence application to be rejected on environmental grounds, the ISA requires evidence of likely environmental harm; for example: 'Prospecting shall not be undertaken if substantial evidence indicates the risk of serious harm to the marine environment' (Regulations on prospecting and exploration for polymetallic sulfides in the Area, Part II, Section 2). That is something of a reversal of the usual 'precautionary principle', whereby evidence of no likely harm is required for an activity to go ahead. The precautionary principle lies at the heart of another UN instrument: the Convention on Biological Diversity (CBD) – for example, Principle 15 of its Rio Declaration states that 'lack of scientific certainty shall not be used as a reason for postponing measures to prevent environmental degradation'.

Some of the current principles followed by the ISA therefore seem to be at odds with those of the UN Convention on Biological Diversity. I came across that issue in 2011 when I led the first Remotely Operated Vehicle dives to hydrothermal vents in the south-west Indian Ocean, where the ISA had already granted the Chinese Ocean Minerals Research Agency (COMRA) an exploration licence for minerals at hydrothermal vents, despite no-one knowing what lived there. Our expedition collected the first samples of vent animals from that region, finding some species previously known from other regions, but also several new species. Although there are certainly more colonies of those new species at other hydrothermal vents elsewhere in the south-west Indian Ocean, for now we don't know where, or how interconnected their populations are. The impacts of developing deep-sea mining at the site we explored will therefore be uncertain until we have that knowledge.

Subsequently, I prepared an application for the hydrothermal vents that we explored to be designated an 'Ecologically or Biological Sensitive Area' (EBSA) under the UN Convention on Biological Diversity. The criteria were clear: the discovery of populations of species not yet known anywhere else on our planet, in a location at potential risk from human activity. The EBSA application was submitted by the UK delegation at the Convention on Biological Diversity meeting for the Indian Ocean in 2012, but perhaps unsurprisingly, the award of EBSA status for those vents was blocked. I understand the reasoning: how could one United Nations organisation (the ISA) award China the rights for mineral exploration activities at that site, and then another United Nations instrument (the Convention on Biological Diversity) designate the same site 'ecologically or biologically sensitive' with regard to human activities? Hopefully the EBSA application for those hydrothermal vents has helped to highlight the current gap between the International Seabed Authority and the UN Convention on Biological Diversity, and encouraged joined-up thinking for the future.

There is a temptation, particularly for those of us who are impatient and restless, to ignore these dry legalities and just focus on exploring the oceans instead. But if we withdraw from what is going on, no longer sharing insights from our exploration where they are relevant for legislators to consider, then we risk seeing a poorly informed future for the oceans. The ISA has now issued twenty-nine mineral exploration licences, covering polymetallic sulfides at hydrothermal vents, manganese nodules on abyssal plains and cobalt crusts on seamounts. For hydrothermal vents, each exploration licence initially covers around 1,000 km of mid-ocean ridge. On the northern Mid-Atlantic Ridge for example, the ISA has awarded four exploration licences to contractors sponsored by France, Russia and Poland for areas south of the Azores. Meanwhile, there are also areas of the ridge that lie within the EEZs of several nations, such as the UK around Ascension, Brazil around St Peter and St Paul Rocks, Portugal around the Azores and Iceland south of its coast. So in just over 160 years since the northern Mid-Atlantic Ridge was first glimpsed as the Telegraph Plateau on Maury's map of 1854, much of its length has now been parcelled up in terms of the mineral resources that it holds.

Northern Mid-Atlantic Ridge

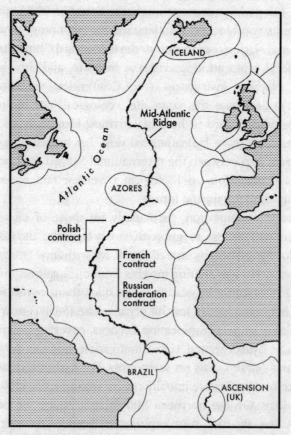

24

Where would you like to explore next?

Most of my work so far has focused on exploring a global jigsaw puzzle of what lives where at hydrothermal vents around the world, and there are still some missing pieces to that puzzle. Having recently investigated what lives at hydrothermal vents in the north Atlantic, the Caribbean, the Southern Ocean and the south-west Indian Ocean, the next place that I hope to explore with colleagues is the southern Mid-Atlantic Ridge. Based on what we have found so far in the neighbouring pieces of the puzzle, we can make some predictions about what we might find at hydrothermal vents in the south Atlantic, to test our current understanding of how life disperses and evolves in the ocean depths. And there's always the chance of a completely unexpected discovery that also transforms our understanding of what's going on in the deep.

The former US Secretary of Defense Donald Rumsfeld famously strayed into the philosophy of exploration when he talked about 'known unknowns' and 'unknown unknowns' in an answer to a reporter's question in 2002. When we are pursuing funding for exploration, we usually highlight the 'known unknowns' that we expect our expeditions to reveal: the gaps in our current understanding of the ocean that we hope to fill from what we find. But leaps in our understanding of the ocean have sometimes come from 'unknown unknowns' encountered during exploration: things we didn't even imagine were going on, or think were possible. There will be plenty of those in the continued exploration of the oceans, but by their nature, we can't really predict where they are and what they might be. It's understandably much harder to secure financial support for ocean exploration from that premise, but it means we have to keep our minds open to the possibilities when we're pursuing the defined objectives of our expeditions, just in case we stumble across 'the next big thing' out there.

I am also looking forward to seeing the start of exploring oceans elsewhere in our solar system, hopefully within my lifetime. Around four billion years ago, Venus may have been similar to Earth, with liquid oceans. But as the sun became brighter, the oceans on that planet would have boiled, and their steam would have created a 'runaway greenhouse' effect in the atmosphere, trapping more and more heat there. Eventually the water vapour split into hydrogen, which was driven off into space by the solar wind – the stream of charged particles beaming from the sun – and the heavier oxygen, which remained to combine with carbon and create a very thick atmosphere dominated by carbon dioxide. Today the atmospheric pressure at the surface of Venus is ninety-three times that on Earth, with a surface temperature around 500 °C, and we need to go further afield to find extraterrestrial oceans.

Mars also may have had a liquid surface ocean more than four billion years ago; analysis of types of water vapour in its very thin atmosphere today is allowing planetary scientists to estimate how much water was stripped away from the young Red Planet by solar winds. The primordial Martian ocean may have covered 19 per cent of the planet's surface, and been 1.6 km deep in places. But today, any water on Mars is probably locked up as ice at the poles or beneath its surface, and the surface atmospheric pressure is now 0.6 per cent that of Earth, with a surface temperature of minus 63 °C. So probably no joy waiting there for ocean explorers either.

Although the outer solar system is cold because of its greater distance from the sun, the huge tides created by massive planets such as Jupiter are forever stretching and squashing the moons that orbit them. That tidal energy is enough to keep water liquid in subsurface oceans on some of those moons, hidden beneath their icy crusts. When NASA's *Galileo* space probe began orbital close encounters with Jupiter's moon Europa in the late 1990s, it measured variations in the moon's magnetic field that could be explained by a liquid ocean sloshing about beneath the frozen surface. It also took pictures of surprising patterns in the icy crust, including sinuous cracks with a very regular wavelength, which could only be explained by tides regularly pulling the crust apart as it floated on a liquid beneath. More recently, the Hubble space telescope has also spotted probable water plumes erupting from the moon's surface. All of this makes Europa a prime candidate for extraterrestrial oceanography, and the search for extraterrestrial life.

Not only does Europa appear to have a hidden ocean, but that ocean is likely to have hydrothermal vents on its sea floor. Europa's closest neighbour, Io, another of Jupiter's moons, is the most volcanically active body in our solar system, as Jupiter's tides stretch and squeeze it, churning its solid body into perpetual volcanic activity. Like Io, the rocky core of Europa will also be stretched and squeezed by Jovian tides, and is therefore very likely to be volcanically active. A liquid ocean and volcanic activity at its sea floor are the ingredients required for hydrothermal vents; so if life began at hydrothermal vents on Earth, then perhaps there could be life on Europa.

Searching for hydrothermal vents and life in a subsurface ocean on Europa presents a new challenge for our ingenuity, but experience and technology gained from exploring the deep oceans on Earth provide the foundation for it. Any robot submarine that manages to get through Europa's icy crust to explore the ocean beneath will have to be autonomous, as direct remote control in real time is not possible given the time delay for signals between Earth and Europa, let alone the challenge of communicating through the ice via a relay station on its surface. But as today's long-range Autonomous Underwater Vehicles continue to improve in their endurance and capability to explore the oceans alone, it is possible that one of their great-grandchildren will one day swim in the ocean on Europa.

Jupiter's moon Callisto may similarly have a subsurface ocean, but its icy crust appears to be much thicker, perhaps reaching 200 kilometres, which makes it a more daunting prospect for exploration. Ganymede may also have an underground salty 'ocean', but one made up of several layers of ice and water sandwiched between its crust and core. Elsewhere, there are hints of possible subsurface oceans on Saturn's moons Mimas and Titan, and Neptune's moon Triton. But the next most promising candidate for exploring oceans elsewhere in our solar system is Saturn's moon Enceladus.

Recent fly-bys of Enceladus by the *Cassini* space probe have found jets of water vapour erupting from the 'tiger-stripe' features of its icy surface, and detected hydrogen in those jets, which suggests a potential energy source for chemosynthetic life. Early estimates put the icy crust of Enceladus at around thirty kilometres thick, but more recent data suggest it could be much thinner, with an ocean up to sixty-five kilometres deep beneath it. At first glance, an ocean sixty-five kilometres deep looks like a new challenge for those building undersea vehicles, far greater than the deepest ocean on Earth. But the gravity on Enceladus is only 1.13 per cent that of Earth, so the pressure at a depth of sixty-five kilometres in an ocean on Enceladus would be equivalent to the pressure at around 750 metres deep in our oceans, which is shallower than the depth reached by Beebe and Barton's bathysphere in the 1930s.

So while our 'blue marble' is unique in having liquid oceans covering its surface, it is not the only ocean world in our solar system. And although our world is a deep-ocean planet from our perspective as air-breathing surface-dwellers, the oceans are actually a very thin layer on our massive ball of rock. In fact, if Earth were the size of an orange, all its liquid water would form a sphere only the size of a pea. Meanwhile, although Enceladus is quite tiny, with a diameter smaller than the length of the UK, it may be proportionally more of a 'water world' than ours. In the same analogy, if Enceladus were the size of an orange, all its liquid water could form a sphere the size of a large grape.

The other water worlds of our solar system should keep ocean explorers busy for centuries to come, in addition to continuing to explore our own deep-ocean world. I don't see any contradiction between exploring the oceans and exploring space: both are exploration of the unknown. Because we can't predict what we'll find and what the benefits will be when uncovering 'unknown unknowns', I don't think we can declare one unknown more worthwhile to explore than another.

25
What is the future for the ocean?

There is a third and final step in exploring the ocean, which is 'using our understanding'. Our exploration of the deep has already revealed wonders to inspire us and shown the impacts that our lives are having on the hidden face of our world. We have developed the technology to explore the furthest reaches of our deep-ocean planet and to share information instantly with people all over the world. How we respond to those discoveries, and use those capabilities, will decide what happens next for the ocean.

When William Beebe and Otis Barton became the first bathynauts in the early 1930s, our global population was just over 2 billion people. Right now it is more than 7.6 billion people. Human population growth represents a success, for example in modern medicine reducing infant mortality, but it also presents a unique challenge for our lifetimes. As development spreads globally, population growth is predicted to level off at about 11 billion people by the end of this century. We therefore have eight decades to figure out how 11 billion of us can live together sustainably, using our planet's resources while supporting greater equality of opportunity for all of us – and a healthy ocean is vital to that outcome.

One possible future for the ocean seems all too familiar: the continued destruction of habitats, and squabbling over resources, for the enrichment of a shrinking proportion of the world's population. But the seeds for another future have also been sown: a growing network of Marine Protected Areas around the world, populations of some previously threatened species brought back from the brink by international agreements, increased awareness and action to tackle pollution and climate change and a realisation of what we can learn from the library of life that the ocean contains.

The final 'using our understanding' step of ocean exploration involves all of us, from politicians creating legislation to each of us making choices in our everyday lives. Our everyday choices are not just the ones we make as consumers, but also where our savings or pensions are invested, and who we vote for. The notion that our own choices are insignificant and of no consequence is a convenient myth that lulls us into inaction: together we can choose a healthy future for the ocean; but if we don't assert our choice then another future will certainly be imposed on us instead, by the vested interests of the past.

Exploring the unknown expands our perspectives, including our perspective on ourselves, and the choices that we make for the future of the ocean are also choices about who we are, or perhaps who we hope to be. As we continue to explore the ocean, will we preserve the biodiversity that we discover, so that we can learn from the ingenuity of nature? Or will we rush to exploit the other resources that we find there, at the risk of losing that biodiversity? Our history contains plenty of examples of the latter, but if we believe we are better than that, and we have grown in wisdom as well as technological capability, then here is the chance to show it. Why we explore the ocean, and what we do with the knowledge that we gain, defines its future and our own.

Index

Page numbers in **bold** refer to figures and tables.

Acknowledgements

This book exists thanks to the creative vision of my editor Briony Gowlett, and her patience with my sporadic progress. I am very grateful to Rosie Collins for her talent as an illustrator, and to the whole team at Hodder & Stoughton for their hard work in making the book a reality, from publicity to proof-reading. I would also like to thank my family for all their support while writing, and I could not have become an "ocean explorer" without their support either, during my time spent away at sea and in the earlier choices of life that led me there.

If you ask any scientist how they ended up pursuing science, most will mention passionate and dedicated teachers who inspired them, and I was fortunate to be taught by several, who kindled my interest in nature at primary school and fueled my fascination with science at secondary school. As a scientist, I have also been fortunate to have superb mentors, especially Prof Paul Tyler, and to work with outstanding research students and colleagues around the world. Finally, I would like to thank the shipmates with whom I have shared voyages, from my first research expeditions as a student to my most recent with the BBC Natural History team. May the wind be at our backs as we continue to explore beyond the blue of our blue planet.

Endnotes

1 John Steinbeck (1966). Let's Go After the Neglected Treasures Beneath the Seas – A plea for equal effort on "inner space" exploration. *Popular Science*, September 1966, pp 84–87.

2 Benjamin Franklin (1786). A Letter from Dr. Benjamin Franklin, to Mr. Alphonsus le Roy, Member of Several Academies, at Paris. Containing Sundry Maritime Observations. *Transactions of the American Philosophical Society*, Volume 2, p 315.

3 Henry Cummings (1874). *A synopsis of the cruise of the U.S.S. "Tuscarora" from the date of her commission to her arrival in San Francisco, Cal, Sept. 2d, 1874.* Cosmopolitan Steam Printing Company, San Francisco. p 28.

4 Matthew Fontaine Maury (1855). *The Physical Geography of the Sea.* Harper & Brothers, New York. p 201.

5 William Beebe (1934). *Half Mile Down.* Harcourt, Brace & Company, New York. pp 99-100.

6 William Beebe (1934). *Half Mile Down.* Harcourt, Brace & Company, New York. p 154.

7 William Beebe (1934). *Half Mile Down.* Harcourt, Brace & Company, New York. pp 185–186.

8 William Beebe (1934). *Half Mile Down.* Harcourt, Brace & Company, New York. p 225.

9 William Beebe (1926). *The Arcturus Adventure.* G. P. Putnam's Sons, New York and London. p 203.

10 Otis Barton (1953). *The World Beneath The Sea.* Thomas Y. Cromwell Company, New York. p 78.

11 Robert Dietz (1959). 1100-Meter Dive in the Bathyscaph *Trieste. Limnology and Oceanography*, Volume 4, Issue 1, p 100.

12 William Beebe (1934). *Half Mile Down*. Harcourt, Brace &
    Company, New York. p 116.

13 Marie Tharp (1999). Connect the dots: mapping the seafloor
    and discovering the mid-ocean ridge. In *Lamont-Doherty Earth
    Observatory: twelve perspectives on the first fifty years 1949–1999*
    (edited by Laurence Lippsett). Lamont-Doherty Observatory
    of Columbia University, Palisades, New York.

14 Allen L. Hammond (1975). Project FAMOUS: Exploring the
    Mid-Atlantic Ridge. *Science*, Volume 187, Issue 4179 (7 March
    1975), p 823.

15 David Perlman (1977). Galapagos Rift Expedition - Astounding
    Undersea Discoveries. *San Francisco Chronicle*, Wednesday 9
    March 1977.

16 Pliny the Elder (c. 77–79 CE) *Naturalis Historia* Book XXXII,
    Chapter 53 (11).

17 Henry Moseley (1880). Deep-sea dredging and life in the deep
    sea. *Nature*, Volume 21 (15 April 1880), p 570.

18 William Carpenter (1871). *Nature*, Volume 4, Issue 84 (8 June
    1871), p 107.

19 Charles Wyville Thomson (1876). The Challenger Expedition.
    Nature, Volume 14, Issue 361 (28 September 1876), p 495.

20 Charles Saxon (1983). *New Yorker*, 21 March 1983.

21 *The Times*, 6 August 1858, p 8.

22 Rudyard Kipling (1893). The Deep Sea Cables. First published
    in *English Illustrated Magazine*, May 1893.